Food, Policy, and Politics

About the Book and Editors

It is a paradox of our time that with many countries of the world suffering severe food shortages and others experiencing a food glut, we have not been able to find the means to solve either problem. In this provocative volume, leading agricultural economists address the issues surrounding food policy with an emphasis on the roles played by the U.S. family farm, markets, government subsidies and foreign aid, Third World politics, and the international economy.

George Horwich is professor of economics and Burton D. Morgan Professor for the Study of Private Enterprise in the Krannert Graduate School of Management, Purdue University. He has published over forty papers in monetary theory and economic policy, including energy policy, and is the author or editor of eight books. His most recent volume, with political scientist David Weimer, is Oil Price Shocks, Market Response, and Contingency Planning. **Gerald J. Lynch** is associate professor of economics at Purdue University and the associate director of the Purdue Center for Economic Education. He is the author of many articles on agriculture, growth, and development.

Food, Policy, and Politics

A Perspective on Agriculture and Development

EDITED BY

George Horwich and Gerald J. Lynch

CRC Press
Taylor & Francis Group
Boca Raton London New York

CRC Press is an imprint of the
Taylor & Francis Group, an **informa** business

First published 1989 by Westview Press, Inc.

Published 2018 by CRC Press
Taylor & Francis Group
6000 Broken Sound Parkway NW, Suite 300
Boca Raton, FL 33487-2742

Visit the Taylor & Francis Web site at
http://www.taylorandfrancis.com

and the CRC Press Web site at
http://www.crcpress.com

Library of Congress Cataloging-in-Publication Data
Food, policy, and politics : a perspective on agriculture and development / edited by George Horwich and Gerald J. Lynch.
p. cm. -- (Westview special studies in agriculture science and policy)
ISBN 0-8133-7725-0
1. Agriculture and state. 2. Agriculture and state--Developing countries. 3. Agriculture--Economic aspects. 4. Agriculture--Economic aspects--Developing countries. 5. Food supply. 6. Food supply--Developing countries. 7. Produce trade. 8. Produce trade--Developing countries. I. Horwich, George. II. Lynch, Gerald J. III. Series.
HD1415.F636 1989
338.1'8'091724--dc19 89-37521
CIP

ISBN 13: 978-0-367-01300-4 (hbk)

Contents

PART 2
THE DOMESTIC PERSPECTIVE

PART 3
SYNTHESIS OF PAPERS

Introduction

George Horwich and Gerald J. Lynch

The papers in this volume address the agricultural problems of both the domestic and international economies. The problems of each are, of course, different, partly because the world contains both developed and developing countries. In the United States and much of the industrial world, huge increases in production have lowered the price of food products so much that many farmers whose families have lived off the land for generations are no longer able to do so. In many developing countries, the problem is just the opposite--not enough food can be grown to support even subsistence levels for the population.

At first blush there appears to be an easy solution--let the surplus nations ship food to the developing nations. In practice, as one of our authors (Mellor) carefully documents, the solution is not so simple. In the surplus nations the individual farmers earn little enough under current conditions; they are certainly not in a position to donate food. The developing countries, on the other hand, do not have the resources to purchase the food. A better solution and ultimately the only viable one is for all nations to solve their agricultural problems individually within the framework of an open international economy.

The papers in this volume discuss the nature of the agricultural problems in both the developed and developing world and suggest specific solutions for each of them. An initial group of papers focuses on the international scene, several emphasizing the experiences of developing countries. A second group analyzes U.S. agricultural policy, generally against the backdrop of the international environment. Although the authors hold a diversity of views on the optimal degree of government involvement in agriculture, there is an approximate (though not unanimous) consensus that government needs to retrench or at least to alter its policies, allowing greater play to agricultural market forces in both developed and developing countries. In much of the industrialized world, including the United States, the appropriate adjustment is one of ending subsidies to agriculture and allowing production to gravitate to its market-determined level (Johnson, White, Butz). Developing countries need to experience the full force of market incentives to increase their food supply, either directly by raising domestic production or indirectly by producing additional exportable goods to fund imports of food on a sustaining basis (Johnson, Shane, Roe).

Beyond this general (though again, not unanimous) consensus, the authors advocate varying degrees of government intervention to facilitate the required adjustments. Some see government as providing primarily the conditions of domestic and international stability within which market forces can operate more or less freely. Several members of this group of authors detail the desired macro policies (Schuh, Shane) while others spell out the role of government in facilitating micro adjustments (Roe). A continuing government presence to equalize rural and urban income, to maintain the family farm, and to overcome market failure in land use is a recurring theme of other contributors (de Janvry, Rausser and Foster, O'Rourke). Mellor calls for a government role in transferring agricultural technology from developed to developing countries, while Dobson and Foster urge restructuring of the U.S. farm credit system. Only one author (O'Rourke), speaking for the Catholic bishops' pastoral letter on American agriculture, strongly supports government-to-government food aid; Paarlberg limits his support to such programs as are properly designed, while Krauss is unconditionally opposed to government aid. A more detailed summary of each paper follows.

Edward Schuh traces major changes in the international economy of the last several decades: the growth of international trade, including agriculture; the emergence of a large and well-integrated international capital market; the shift from fixed to bloc-floating exchange rates; and the appearance of a significant degree of monetary instability. In this environment agriculture bears the major impact of changes in domestic monetary and fiscal policies, which directly affect capital flows and exchange rates more than they do interest rates, imposing increased instability in agricultural demand. Agriculture also bears the brunt of adjustments in exchange rates, whatever their cause. Thus the explosion of third-world debt has required countries like Brazil to devalue their currencies in order to service their debt through increased exports. Exports, including food, of advanced countries have suffered proportionately. In the United States agricultural policy should avoid pricing American crops out of the world market. Among advanced countries, monetary and fiscal policies should be coordinated to avoid large differences in real interest rates.

Kelley White, in the next paper, characterizes world agricultural markets by citing their increased internationalization; their "thinness" due to domination by relatively few producing countries (and within these countries by a few large firms); their increased volatility owing to the use of more marginal land in more variable climates and to domestic policies that shift internal adjustments onto the world market; and market imperfections, which, because of improved communication and continued freedom of entry, have been reduced, but because of increased domestic protection and state trading, have been increased. White details the U.S. role as both a major exporter and importer of agricultural products and, because of its domestic support policies, as the world's major stockpiler and source of price stability, particularly with respect to downward movements. An examination of events in the 1970s and 1980s indicates that the United States is unlikely to regain its former dominance in export markets.

John Mellor offers a historical perspective on the world food situation, stressing its diversity. Although many (but not all) third world countries suffer from a food deficit, much of the rest of the world is in a food glut. A short-run solution appears to be food trade from the surplus to the deficit countries. Yet, most of the less developed countries have little of value to trade and almost no foreign exchange. Most developing countries are single crop exporters and the increasing variability of food prices since the 1970s has damaged their earnings capability.

A superior solution for the LDC's is to develop an agricultural strategy of development. Mellor believes that such a strategy should employ more agricultural technology. This is particularly important since the amount of land devoted to agriculture has declined. In the past two decades, almost all of the growth in food production has come from increases in productivity. He suggests that the developed world must make more financial and technical aid available to the developing countries.

Mathew Shane notes that intervention by the governments of the developing countries has not always been productive. Shane characterizes two national patterns of economic and agricultural development. One set of countries implicitly tax their agriculture and impose disincentives on industrial production directed both to domestic use and export. They thus lack foreign exchange for food imports and are "food crisis" countries, borrowing and relying on others for food aid. A second category of countries experiences dynamic growth, including exports adequate to pay for required imports of food. The governments of crisis countries are shown to have established artificially low domestic food prices and overvalued currencies resulting in (a) declining exports and foreign trade relative to gross domestic product and (b) stagnant per capita incomes. The growth countries have an opposite pattern; their foreign debt, moreover, has increased more rapidly than that of the crisis countries, but not in relation to exports or GDP.

Terry Roe describes the role of government in developing countries in greater detail, examining the various means used by governments to accelerate the transfer of resources out of agriculture. Many, if not most, interventions for this purpose are inefficient in that they distort the price and incentive structure of both the agricultural and industrial sectors. Among such interventions are direct restrictions on foreign trade, such as limits on agricultural exports, and, in order to protect domestic industry, on nonagricultural imports. Domestic interventions take the form of artificially low food prices for urban consumers; agriculture is compensated for the resulting negative impacts by government subsidies to agricultural inputs and by marketing enterprises operated by the government, typically at a loss. On the macro level, the interventions create an overvalued currency, a foreign trade deficit, a domestic fiscal deficit, growing indebtedness, excessive growth of the money supply, and, predictably, a sharply reduced ability to adapt to shocks in world markets.

In explaining why these policies are undertaken, Roe

finds that to some degree they are the outcome of political pressures by self-seeking groups. They are also the result of policy mistakes due largely to the decentralization of the policy process and they are the aftermath of policies for which the social costs were initially small but have become large. Meanwhile, the policies are politically irreversible owing to high costs of disengagement that tend to fall disproportionately on the poor. Roe urges removal of all interventions that give rise to the above distortions. Government programs should be redesigned and implemented only where markets clearly fail, public enterprises should be divested by equitable means while natural monopolies should be organized so as to minimize cost and price, distorting (unequal) tariff and tax rates on imports and exports should be removed, and low income households should be compensated for the adjustment costs by carefully targeted programs.

Melvyn Krauss, long an iconoclast in regard to the role of U.S. aid to developing countries, argues that such aid has been no more beneficial to the people of developing countries than have the domestic government policies of those countries. Krauss asserts that food aid and technical assistance from the United States in the past have actually done more harm than good. Most of the aid does not actually reach the poor; moreover, as British economist P. T. Bauer has documented, aid typically goes to the governments of the developing countries who then use it, in whatever form it comes, to subsidize the elite and buy votes.

Krauss cites Bangladesh, where food that is donated is sold at below market prices. But purchasers must have a ration card and only a select few receive them. Krauss believes that in the recent famine in Ethiopia, too little food again reached the poor. In general, even if the food does reach them, it tends to free up resources for domestic production of goods that could be imported more cheaply. All too often the domestic production is undertaken by governmental enterprises that are economically inefficient. In addition, the gift of food results in lower prices for agricultural products in the receiving country, discouraging domestic production.

If food aid is so damaging, Krauss asks, why do we have it? He concludes that the recipient governments see food aid as a way of facilitating industrialization that, in the hands of the government, proceeds inefficiently. But the result is a zero-sum-game where farmers in developed countries experience a high demand for their products at the expense of less developed farmers. He makes the same indictment of technical assistance--that it often crowds out the private sector in the less developed countries.

Don Paarlberg responds that a balanced perspective requires that the successes as well as the failures of food aid be cited. He recalls the success of food aid in fighting the famine in the Bihar Province of India in 1966-67. A prerequisite of receiving food in that particular instance was that price ceilings be lifted on agricultural products. By 1982-83, wheat production had trebled in the province. Paarlberg's basic point is that despite some admitted failures, food aid programs are not inherently flawed. The failures are administrative in nature and can be avoided.

Alain de Janvry suggests in the final paper on the international perspective that land reform can also be beneficial under the right circumstances. de Janvry discusses the farm size and tenure arrangements that maximize farm productivity. Land reform, as carried out in less developed countries, usually redistributes land from large single-owner farms, where there is a great deal of sharecropping, to small family farms. de Janvry finds that under a restrictive set of assumptions, the family farm can be more productive than the large sharecropper farms. In particular, if there is absentee ownership of the large farms, and if there is not an effective labor market for women and children, family farms may be more productive. Productivity may also increase when farmers go from a coercive relationship as sharecroppers to an unsupervised role as land owners on the family farm where they are more likely to exert a maximum effort.

On the other hand, there are factors in land reform that work in the opposite direction. If larger farms have access to lower credit costs, if there is continuous land-saving technological change, or if there is an efficient rental market for land, land reform may reduce output and productivity. Although de Janvry is aware that the conditions for the economic success of land reform are very restrictive, he concludes that it is still often a worthwhile form of social policy in the rural sector.

The next set of papers focuses on the domestic farm problem. Like other developed countries, overproduction in the United States is due to the inability of government to control market forces that counterveil its policies. This point is developed in an overview of U.S. agriculture by D. Gale Johnson. Johnson cites the existence of substantial excess capacity as the central problem in U.S. agriculture today. Although the Food Security Act of 1985 permits market prices to decline gradually to what may be market-clearing levels, the act continues numerous subsidies that run counter to market forces. In particular, a new feature permits farmers to repay price support loans at much less than the original amount of the loan. Johnson sees little prospect that U.S. exports can recover their former share of world markets, even under lower exchange rate and price support levels. He finds our high support prices and export subsidies inimical to free world trade and particularly harmful to the market-based export position of developing countries.

Given that setting, Gordon Rausser and William Foster discuss what a coherent policy for U.S. agriculture would entail. They offer a "logically consistent" set of agricultural policies to achieve a given set of goals. Their coherent agricultural policy urges the elimination of commodity-specific programs; the continuation of low-income food subsidies and the system of food inspection and regulations; the maintenance of agricultural research while attempting to impose the costs more equitably on beneficiaries, to reduce government-supported research that may result in larger, more capital-intensive farms, and to publicize more widely the costs and resulting benefits; the development of a tax structure that achieves greater neutrality as between agriculture and nonagricultural sectors and removes incentives to investment in farmland and in capital-intensive farm production; direct income transfers to farmers based on designated income and farm-asset levels; a variable credit policy designed to dampen the swings in agricultural investment; flexible storage rules designed to moderate price swings; adoption of greater incentives to retire erodible land and decrease use and runoff of chemicals; and the use of economic retaliation, such as export subsidies and cartel price-fixing in response to subsidies to inefficient producers by other countries. This last recommendation, as Rausser and Foster are aware, is one that most economists, including those at the conference, have considerable difficulty accepting.

Perhaps the most visible side of the current agricultural problem is the relationship between farmers and financial institutions. In the mid-1980s, the press heralded the showdowns, often violent, between farmers attempting to save their land and bankers trying to foreclose. William Dobson and Freddie Barnard explore a less well known facet of that problem. The farmers were not the only ones facing financial distress. The lending financial institutions also saw their assets erode in value; many of them came perilously close to failing.

Dobson and Barnard find the basic source of the problems of the Farm Credit System (FCS) in the unanticipated economic recessions of 1979-1982. To compound the problem, in 1984 the FCS initiated a study of future economic conditions that offered a considerably more optimistic picture of the level and growth of U.S. agricultural exports than actually materialized. The FCS made loans based on this optimistic outlook, acquiring assets they might not otherwise have taken on. Dobson and Barnard are quick to note that the loan portfolio of the FCS does not appear to be any riskier than that of the private sector. Although private financial institutions also suffered tremendous losses on their farm loans, however, their portfolios were generally more diversified than those of the FCS.

In response to the growing problems of the FCS, Congress initiated legislation in 1985 to guarantee its loans. The market reacted quickly to this assurance; the yield spread between FCA debt instruments and Treasury bills dropped immediately and the exodus of depositors and borrowers from the system ceased. But if the FCS is to evolve into a viable credit system over the next few years, it must reduce the risk of its portfolio by diversifying. Dobson and Barnard observe that such diversification runs counter to the very rationale for the existence of the FCS, but that may be its only hope for survival.

Another widely publicized viewpoint on the farm problem is contained in the Catholic bishop's pastoral draft statement on American agriculture. Bishop Edward O'Rourke, speaking for the bishops, argues that U.S. government policy has kept food prices low for consumers, while pressuring farmers to increase output and reduce cost. This has led to a displacement of farm labor, an expansion of farm size, disregard for soil and water conservation, underpayment of farm workers, and opposition to farm worker unionization. A "preferential option for the poor" should be the central thrust of any governmental policy toward agriculture. Policy should be directed at increasing all of the following: farm ownership by minorities, the number of small or medium-size farms, part-time farming, the general employment level in farming, and food aid (unmotivated by politics) to Third World countries. U.S. support for farmers should be directed to those truly in need (including those experiencing debt crises), government-supported research should focus on raising the productivity of small and medium-sized farms, and policies should be adopted toward promoting conservation while allocating the resulting costs more widely.

In his comment on Bishop O'Rourke's paper, George Horwich argues that the Catholic bishops' assessment of American agriculture lacks both historical and economic perspective. The bishops fail to see the continuing exodus from the land, the declining use of land, and the rising scale of activity--all dictated by technology--that has characterized all of U.S. history. Attempts to stem this basic economic tide cannot ultimately succeed. He disputes the bishops' contention that farm policy has benefited consumers at the expense of farmers; U.S. policy has tended to support prices at higher than market levels, harming consumers and maintaining a greater farm population, most of whom are not small farmers. Horwich also disputes the relevance of the "preferential option of the poor" to the farm population, most of whom, despite their current financial difficulties, are neither without "voice or choice," the characteristics of the poor attributed by the bishops. He also believes unionization tends to hasten the decline of industries--something agriculture does not need, and questions the desirability and feasibility of encouraging minority groups to enter farming when, for so many, their economic progress is measured by the distance they have traveled from their rural agricultural roots.

The final two papers in the volume return to the thread that runs through the papers in the volume. Earl Butz, in a personal reminiscence, discusses the decline of the family farm, a result primarily of the increased use of capital in agriculture. He is thankful that there was no federal program that would have discouraged him from leaving his family's farm. Nor does he see any sign that the movement off the land has slowed, even though the fraction of the American population living on farms has declined in this century from 40 percent to 2.2 percent and the number of people fed per farm worker has risen from 5 to 75. In spite of individual cases of distress that attract so much media attention, Butz feels the flight from the farm should not be impeded. Farm income will not be enhanced by restrictive government controls, rising costs, unrealistic pricing, or increased transfer payments from the U.S. treasury. For Butz the key words to success on the farm are efficiency, competitive pricing, and a favorable macro environment. In the latter respect the conference ended on the note it began with in the paper by Edward Schuh--one of the best things government can do for the farmer is to create a favorable macro environment and leave him to his own choices.

Gerald Lynch synthesizes all the papers in a comment that focuses on why farmers receive subsidies. He suggests that efficiency is not the basic reason. If the market felt that some particular producer was the low cost producer of a good for which it was willing to pay, there would be no need for an additional payment. The very existence of a subsidy usually signals that there is an intervention in normal resource allocation.

The next question then is, why is the farm sector singled out for subsidies? Farmers are not the only group that has suffered financial difficulties over time. They are certainly the group, however, that has received the most attention. No one seems to lament the loss of buggy whip companies because their demise was evidence of a structural change in the economy that was viewed positively. To Lynch there is no apparent reason why the farm sector receives so much more assistance than the rest of the economy unless it is the romantic notion of the importance of the family farm.

The last comment addresses one of the major problems in assessing the efficiency of farm programs. As many of the papers note, there is an intermingling of economics and emotionalism in addressing agricultural issues both at home and abroad. Although this two-edged aspect of the question makes it difficult to draw conclusions about efficiency, the papers in this volume try to ask the right questions.

PART ONE

International Aspects

1

Impact of the International Macro Environment on the World Food Situation

G. Edward Schuh

An assessment of the macroeconomic environment is the key to understanding the international food and agriculture economy. Moreover, despite the importance of macro issues, they are poorly understood and we end up making significant errors in policy because of the lack of understanding.

In focusing on the macroeconomic environment, I will limit myself to monetary and fiscal policies implemented by national governments, the international capital market, and the exchange rate or value of a nation's currency either as established by governmental decree or in the markets for foreign exchange. Monetary policy, of course, refers to actions that governments take vis-a-vis the amount of money in the economy--in our case, actions taken by the Federal Reserve. Fiscal policy, on the other hand, refers to the tax and expenditure policies of the government.

This paper is divided into three parts. First, I will provide some background on the international economy so as to establish a proper setting for what is to follow. Next I will discuss the implications of the kind of international economy we now have. And finally, I will take stock of our present circumstances and their implications for the world food situation. I will then offer some concluding comments.

In these last twenty years, the international economy has undergone what can only be described as dramatic changes. These changes have significantly altered the ways that national economies relate to each other; they have modified the ways in which monetary and fiscal policies affect national economies; and they have created linkages among national economies that are very different than what we had in the past. I want to review briefly these changes since they are the key to understanding how we must think differently today about policy and development.

The first thing we need to recognize is that the world has experienced an almost steady growth in dependence on international trade throughout the post-World War II period. During this time, international trade has been growing faster than world GNP--five years (1952, 1958, and 1980-82) excepted. As a result, our respective economies have become increasingly open to international economic forces. The corollary, of course, is that our national economies are increasingly beyond the reach of national policies. This has been an enormous source of frustration to individuals and interest groups, since policies that in the past had well identified effects on the economy no longer do. U.S. agricultural policies are a perfect example.

Agriculture is no exception to the general growth of dependence on trade. Even though total agricultural trade has been declining significantly in the early 1980's, agriculture is still the most internationally integrated sector of our economy. Almost every country either imports or exports food and agricultural commodities, and many do both. This is a very significant development.

The second major change in the international economy has been the emergence of a well-integrated international capital market. This market began with the emergence of the Euro-dollar market in the 1960's and was followed by the Euro-currency market and later the flood of petro-dollars as OPEC raised the price of petroleum, which was transacted in dollar terms. Recall that commercial banks were enjoined to recycle those dollars, which they did to a fault, lending astronomical amounts mainly to third-world consuming countries. Today we refer to the consequences of that episode as the international debt crisis.

Today, international capital markets are enormous. In 1984, the last year for which we have complete data, total international financial flows among countries were on the order of $42 trillion. Total international trade flows, on the other hand, were only $2 trillion. The capital markets are unprecedented in history and, given the present configuration of the international economy, are the driving force in it. Yet, our thinking about these markets tends to be narrowly focused on the international debt crisis. We need to take a broader view.

The third major development in the international economy was the shift from the Old Bretton Woods fixed-exchange rate system to the system of bloc-floating exchange rates in 1973. As we will see, this shift changed the way in which monetary and fiscal policies affect national economies, with enormous significance for agriculture.

The final change was the emergence of a great deal of monetary instability starting in about 1968. Unfortunately, we understand this increased instability only poorly. But in light of the other developments in the international economy, monetary instability has special significance for agriculture, as we shall see momentarily.

Implications of These Developments

By contrast, in the 1950's and early 1960's, when fixed exchange rates prevailed and the international capital market was limited, monetary and fiscal policy had very little impact on agriculture. Such impact as it did have was primarily through the labor market, since the rate of out-migration was determined in large part by the level of unemployment in the domestic economy. The effect of monetary and fiscal policy fell primarily on interest-rate sensitive sectors such as housing, construction, and durable goods.

In today's world of floating exchange rates and a well-integrated capital market, agriculture is no longer isolated from these influences. In fact, as a trade sector, agriculture in most countries now bears the burden of adjustment to changes in monetary and fiscal policies. The reason for this is that these policies generally can now have only modest effects on interest rates. Instead, monetary and fiscal actions affect the economy largely through changes in the exchange rate or the value of a nation's currency. Hence, tight money policies, for example, draw capital from abroad, leading to a stronger currency, which attracts cheaper imports and curtails exports. Easier monetary policies, on the other hand, drive capital out of the country, weakening the currency, reducing imports, and raising exports.

This sequence of events creates a significantly different world for agriculture. Rather than facing a stable demand, as it did generally prior to the changes in the world economy, agriculture now faces a relatively unstable demand due proximately to the instability of exchange rates.

The experience of U. S. agriculture since the early 1970's dramatizes this new context for international agriculture. The large and sustained fall in the value of the dollar during the 1970's contributed importantly to the U. S. export boom of that period. (Rapid and sustained expansion of the international economy also played a significant role.) The unprecedented rise in the value of the dollar in the 1980's has had just the opposite effect, further complicated by rigid commodity programs which limited the adjustment in prices to the changed international conditions. The resulting stress on U. S. agriculture has been amply documented. This time the problem was exacerbated by a stagnant world economy.

The changes in exchange rates have been long swings. But in the new world economy, there has also been a great deal of short-term instability in exchange rates, causing a significant increase in short-term instability in commodity markets. This brings us to the second implication of the changed international economy: the strong linkage we now have between financial markets and commodity markets. From a monetary standpoint, that is due to the emergence of increased monetary instability. But the problem is exacerbated, as Professor D. Gale Johnson has reminded us[1], by trade barriers that limit adjustment in individual countries and thus divert the shocks into the international economy. The diversion imposes large adjustments on relatively open agricultures such as that of the United States.

What I have described thus far is the core of the new analytical framework necessary to understand today's commodity markets and the particular nature of the international food and agriculture sector. But two additional related issues are very important. The first is developments in the petroleum sector in the last fifteen years. The significance of these developments derives in part from the fact that the bulk of the transactions in petroleum is in dollar terms, and in part from errors and large swings in U. S. petroleum policy.

Let's go back to the 1970's. An important reason for the sustained weakness of the U. S. dollar in that decade is what can only be described as a misguided energy policy. The regulation of the domestic petroleum industry in effect subsidized the importation of high-priced OPEC oil while effectively taxing domestic oil production. Our petroleum import bill consequently burgeoned, reducing the value of the dollar. The weaker dollar contributed to the agricultural export boom of that period. What was terrible economic policy for the economy as a whole was clearly good for agriculture.

One of the first things President Reagan did when he came into office was to deregulate the domestic petroleum industry. Deregulation was followed by reduced petroleum use, increased output, and a sharply reduced petroleum import bill. Out of these forces came a dollar in the 1980's much stronger than it otherwise would have been. (Recall that oil deregulation coincided with the very tight monetary policies of that era.) With the rise in the dollar came a collapse in agricultural commodity markets.

This experience with the petroleum sector illustrates how internationally interdependent the food and agriculture sector has become. But the interdependence doesn't stop there, and this is the second set of issues related to the core of the new analytical model we need to use. There is further interdependence that comes about through what I call third-country effects of exchange rate realignments. These effects arise from the fact that rather than having a completely flexible exchange rate regime, we have instead a bloc-flexible or bloc-floating exchange rate system. We have a system in which the major currencies of the world, such as the U. S. dollar, the British pound, the Japanese yen, and the German Deutschemark, float relative to each other, while simultaneously many developing countries fix the value of their currencies in terms of the value of one of the major currencies. Consequently, as the major currencies change relative to one another, they carry along the currencies tied to them. The result is a great deal of implicit flexibility in the system. Moreover, rather large shocks can be imposed on the economies of the smaller countries as the major currencies fluctuate relative to one another. Relationships such as these contributed importantly to the problems of Brazil and Mexico.

A final myopic tendency, overlooking larger interdependence, is the habit of viewing international economic relations largely in terms of trade flows. When we think of imbalances, for example, we tend to think of imbalances on the trade account. We talk about trade deficits and trade surpluses. That was appropriate in the 1950's and for most of the 1960's. But in today's world, the international capital market is far more important than is international trade in determining the value of national currencies. More fundamentally, the trade and capital accounts of our exterior relations are interrelated and thus impose a dual constraint (or dual opportunity, if you will) on our economy and on our policymakers. By definition, these two accounts have to offset each other.

Consider the following highly simplified example. Suppose Brazil has acquired a lot of foreign debt exclusively by borrowing from the United States, and that this debt needs to be serviced and eventually repaid. In order to accommodate the U. S., Brazil will have to run a surplus in its trade accounts. But given that the international accounts have to balance overall, the United States will simultaneously have to run a trade deficit.

This dual constraint has still another implication. If the United States is to run a trade deficit, it also has to be a net borrower on its international capital accounts. Hence the net inflow of capital into the United States is an important consequence of the new configuration of the international economy. (I hasten to add, however, that this does not justify our tremendous borrowing from abroad to finance our large budget deficit. That is another matter.)

Let us look now at some of the implications of this dual constraint. First, we have international burden sharing of perhaps the most important kind. We may wish that the developing countries with all their problems would go away, but they won't. If we want our loans to be repaid, we must be willing to accept imports from these countries and make the parallel adjustments in our own economy.

Second, for the first time in our modern history, important groups in our economy have a vested interest in trade liberalization. If the bankers are to have their loans repaid and/or serviced, we must be willing to accept imports from the debtor countries. The bankers are a powerful lobby and have helped this nation resist protectionist pressures rather well.

Finally, the dual constraints become less binding and less severe in their effects the more rapidly the international economy grows. That is what is essentially behind the so-called Baker Plan, which proposes to deal with the international debt crisis by having commercial banks and multilateral lending institutions lend additional funds to those countries in difficulty—subject to major policy reforms—so as to reactivate economic growth. It is also why helping to revitalize economic growth in the developing countries and in Western Europe is so vital to our own economic performance.

Where Do We Stand Now?

Both the United States and the rest of the world have gone through an enormous adjustment period as we have attempted to respond to the changed international realities I have described above. It is in fact a marvel that we have done as well as we have, given the size of the shocks and the new configuration of the international economy we have had to adjust to.

Where do we now stand? First, we are in a situation in which food supplies are in what is variously described as a surplus situation. Stocks have built up rapidly around the world, especially in the United States, prices are low, and farmers are suffering very low incomes. Some analysts have described it as "A World Awash in Grain."

There are two general comments I would like to make on this situation. First, to describe the world as in a surplus food situation is an exaggeration. The so-called surplus is in large part a consequence of U. S. and European Community agricultural policies. There are numerous countries in which national food supplies are not keeping up with population growth. And there are hundreds of millions of malnourished people in countries that have attained food self-sufficiency. India is an outstanding example.

Second, in many respects we are now reaping our own whirlwind. The combination of high grain prices legislated in the 1981 farm bill, the legislated escalation of those prices, the lack of downward flexibility permitted in those prices, and the unprecedented rise in the value of the dollar had very significant impacts on both U. S. and world agriculture. Had we attempted deliberately to design a policy that would reduce our market shares, we could not have done a better job.

One of our perverse policies was providing producers in other countries a strong impetus to increase their production. Given our relative importance in the wheat, corn, and soybean markets, our high prices—especially as reflected in other currencies through the strong dollar--offered strong incentives to producers in various countries to increase their production. And they have.

These incentives have been reduced by the 1985 farm legislation, which has lowered the loan levels, and by the decline in the value of the dollar. We need to keep in mind, however, that these adjustments do not take place immediately. Nor are these the only factors at work in the international economy.

Also operating are the policy reforms taking place in many developing countries. Interestingly enough, these reforms are often a byproduct of the effort to cope with the international debt problems. Debtor countries need to increase their exports and reduce their imports to create the necessary trade surplus. Devaluing their currency is an important means of bringing about these changes. (An overvalued currency is the most common means by which developing countries in the past have discriminated against their producers.)

These policy reforms will stimulate food production in many countries. Whether this increase in output will offset the contrary effects of the 1985 farm legislation (with its decline in loan levels) and the fall in the value of the dollar (which should lower prices in terms of other currencies) is an open question.

My discussion has emphasized the role of the exchange rate in both the evaluation of U. S. policy and the policies of the developing countries. This raises the question of the outlook for the U. S. dollar. In taking stock of this issue, it is important to recognize that the very strong dollar of the early 1980's was a consequence of highly contradictory monetary and fiscal policies in the United States, and the fact that the countries of Western Europe and Japan pursued diametrically opposite policies.

In the United States, we have had large budget deficits, which tend to be stimulative in their own right. But Mr. Volcker and the Federal Reserve have until recently brought inflation under control by pursuing very tight monetary policies. The result has been interest rates that in real terms were the highest on record. This was a powerful attraction drawing capital into the country, and this capital inflow made for a strong dollar.

But exacerbating this problem were opposite macro policies in Western Europe and Japan, who pursued very conservative fiscal policies and rather easy monetary policies. This created relatively low interest rates in these regions, further stimulating a flow of capital into the United States and the rise in the dollar.

What can we expect in the future? The dollar has now fallen some 30 percent over the last year. This seems to be due to two factors. First, inflation has been squeezed out of the U. S. economy and inflation premia have declined significantly. This has led to a significant decline in both nominal and real interest rates in this country.

In addition, there has been a great deal of coordination of monetary and fiscal policies among the industrialized countries in the last year or so. We have not reached Nirvana by any means, but policies have converged enough to reduce the interest-rate differentials between the U. S. and other countries. This convergence of interest rates, in my view, has played a larger role in the fall of the dollar than has the coordinated sale of dollars by the group known as G-5 (which includes Britain, France, Japan, West Germany, and the U. S.). These foreign exchange markets are simply too big to enable coordinated dumping to exert more than a marginal influence.

I must confess, however, to having little faith that macroeconomic policies will continue to be coordinated. Despite the rhetoric coming from the recent Tokyo Summit, I suspect that under pressure each country will pursue what it believes to be in its short-run interest. Hence, I don't think that anybody can predict the future of the dollar with any degree of certainty. And that is basically because we can't predict what governments will do.

There is one last set of issues I want to discuss before turning to my conclusions. This has to do with the large multi-year swings in the value of national currencies. Underlying these long swings, of course, is the considerable amount of instability in commodity markets induced by monetary instability. The instability of currencies imposes rather large external shocks on the agricultural sectors of many countries. The instability of commodities creates a great deal of risk and uncertainty, which most developing countries cannot transfer to other parts of the economy by institutions such as futures markets.

One consequence of these two phenomena is worldwide pressure for protectionism. Rather than experiencing and adjusting to instability and uncertainty, many countries choose instead to isolate their economies from the world order. To the extent that they do, we are all net losers.

Concluding Comments

I would like to conclude with two comments. First, the vitality and health of the international food and agriculture economy depend importantly on the development of a more stable international monetary environment. This is not the place to discuss how this might be accomplished, but it is important to recognize that the issue should be high on our policy agenda.

Second, we need to recognize how important is economic adjustment in the present world. Neither U. S. farmers nor farmers in any other country can expect to produce at capacity year in and year out. Farmers must respond to changing economic conditions, as do producers in other sectors of the economy, and adjust their production from year to year. Otherwise, both our domestic and international systems will cease to function. This will be in the interest of none of us.

In this paper, I hope I have illustrated how macroeconomic policies--monetary, fiscal, and exchange rate--are at the nexus of the world food situation. In today's world they are far more important than most domestic agricultural policies, in part because the exchange rate is the most important price in an economy, and in part because of the relative importance of trade. The only thing that comes close to these policies in relative importance is trade policy. But even that is intertwined with the macroeconomic policies just reviewed.

Notes

1 "World Agriculture, Commodity Policy, and Price Variability," American Journal of Agricultural Economics, 57(5), December 1975, pp. 823-28.

2

Trade and Agriculture

T. Kelley White

There is an old Chinese proverb which says, "May my enemies live in interesting times." The past 15 to 25 years have certainly been interesting times as world and U.S. agriculture have undergone radical changes. One of the most important of these has been the growing importance of international trade in agricultural commodities. The United States has been intimately involved as trade became a more important component of world agriculture and the growing interdependence between U.S. and foreign agriculture has increased the dependence of U.S. agriculture on world markets. This dependence has been viewed alternately with euphoria and dismay by both farmers and policymakers as changing conditions and U.S. policies have together brought about sudden and significant changes.

The objective of this paper is to place U.S. agriculture in historical perspective with particular emphasis on international trade and the condition of world markets. The paper is divided into five sections. The first section describes a few important and relevant characteristics of world agricultural markets. The second section discusses the U.S. role in world markets; the third section, the role of international trade in U.S. agriculture. This is followed by a brief discussion of the specific conditions in world markets that have brought us to the current state of agriculture in this country. The final section identifies what appear to be emerging trends in both world markets and in U.S. policies and their implications for the future.

Characteristics of World Agricultural Markets

World agricultural markets can be described in terms of a wide variety of structural and performance characteristics. I will focus on three: (1) the thinness of most international agricultural markets; (2) the volatility of such markets; (3) the imperfection of these markets. This section will conclude with a comment on the effect of these three characteristics on price responsiveness in the international setting.

Thinness of Markets

The importance of international trade in world agriculture is a recent phenomenon, as indicated in Figure 1 (Figures at the end of chapter). In 1960, less than 10 percent of the world's production of wheat and coarse grains entered into international trade. After rising to almost 14 percent in 1965, the proportion fell below 10 percent in 1968. This decline was followed by a prolonged and relatively steady increase until the proportion peaked in 1980 at approximately 17 percent. For a number of reasons that will be discussed later, the proportion of total grain production entering into world trade has declined since 1980. Meanwhile, we take note of the sustained growth in trade during the decade of the seventies and the fact that even at the 1980 peak, less than 20 percent of world grain production entered into world trade. This is a relatively moderate percentage that enables large producers to exert significant impact on the price, a characteristic of thin markets.

On an individual commodity basis, world trade as a proportion of world production in 1980 was 22 percent for wheat, 20 percent for corn, 33 percent for soybeans, 33 percent for sugar, and 7.5 percent for beef. In view of these percentages, the more important individual producing countries are large relative to the volume of world trade. For example, in 1980 a 10 percent change in U.S. corn production equated to a 30 percent change in the world trade in corn, and a 10 percent change in China's rice production amounted to nearly a 100 percent change in the world rice trade.

World agricultural markets can also be said to be thin in a second sense. While there are more than 160 countries in the world, agricultural trade is dominated by flows among a relatively few. For example, during the early 1980s, the top five exporters accounted for 94 percent of wheat exports and 93 percent of corn exports, while the top three exporters accounted for 95 percent of soybean exports. On the import side, the top seven importers of wheat accounted for 56 percent of wheat imports; the top six importers of corn, 50 percent; and the top two importers of soybeans, 60 percent of the respective import totals.

The relatively small number of decisionmakers within each country participating directly in world agriculture is also relevant to the structure of the market. Even in the U.S., where trade is conducted by private-sector firms, the bulk of agricultural trade is conducted by a relatively few very large firms. In most of the less developed countries, in all of the centrally planned economies, and in many of the developed market economies, imports and exports of agricultural commodities are handled totally, or in large part, by government agencies or state-sanctioned monopolies.

Volatility of World Markets

Much concern has been expressed about the volatility of world agricultural markets and the resulting impact on U.S. commodity markets, particularly in view of our increased international participation. An earlier paper in this conference by G. Edward Schuh focused on macroeconomic conditions and policies as the source of volatility. Increased production variability has also been cited as a source of volatility as world production has expanded into areas with more marginal land and more variable climate. Table 1 (Tables at the end of chapter) presents measures of variability in agricultural production for selected countries and regions for 1961-72 and 1972-83. These data suggest that production variability in individual countries and selected regions of the world has increased. The data should be treated with caution, however, since measures of variability are very sensitive to the particular time periods selected. Comparing the measure of production variability for the world with that of individual countries and regions indicates that as one aggregates over countries (and commodities, as well), the variability of production declines. The important implication is that aggregate world production variability is small enough so that if world markets were allowed to work, the shocks originating from changes in production could be handled without major price disruptions.

Demand is also an important source of volatility in world agricultural markets. The very rapid growth in agricultural trade during the decade of the seventies was associated with a period of rapid income growth--especially in the less developed and middle income countries--accompanied by rapid monetary expansion, inflation, and expansion in international credit. Likewise, the slowdown in international trade in the first half of the 1980s has been associated with a global recession, contraction in credit availability, and a decline in inflation and international liquidity. Demand shocks tend to be of longer duration than production shocks and more pervasive across countries. They are thus more likely to induce significant price changes in world markets. However, attempts by individual countries to protect their individual producers and consumers from demand or supply shocks have the effect of shifting variability to the world market and thus exacerbating the price effects of shocks (McCalla and Josling).

Imperfection of Markets

International markets for most agricultural commodities certainly fall far short of meeting the conditions for perfectly competitive market structure. As noted above, the limited number of entities participating in world commodity markets and the large size of individual countries relative to the total market could enable individual buyers or sellers to influence price. This influence is constrained, however, by relatively free entry into markets, even though there are resource, climatic, and technological forces that preclude or impede entry of some countries into some markets. But there are no really effective institutional barriers to entry over the long run. This has been clearly demonstrated by every attempt to maintain international agricultural cartels.

Knowledge of market conditions is not perfect in international markets, as it is not in domestic markets. However, recent technological developments in communication have made world market information more quickly and more uniformly available than ever before.

The most important imperfections in world agricultural markets, at least in the short run, derive from the various policies and practices of both importing and exporting countries in their attempt to shelter domestic producers and consumers from changing conditions in the world market. These policies essentially distort world price signals. The pervasiveness of this practice is indicated in Table 2 (Tables at the end of chapter), which presents, for corn and wheat, the percentage of imports going to countries that are basically free traders, as opposed to those that are state traders or that use variable levies to filter out international price signals. World markets are, of course, further distorted by the use of implicit and explicit export subsidies and taxes.

While world agricultural markets may not be perfectly competitive, there is considerable evidence that they are price responsive, especially in the longer term. The recent loss of market share by the U.S. due to the relatively high prices of U.S. commodities (resulting from the high value of the dollar and domestic price supports) is a very real indication of price responsiveness in world markets. The rapid growth in agricultural production in the European Economic Community and in the world generally in the late seventies in response to high world prices support this interpretation. Economists are in general agreement that world markets are more elastic than individual country markets and that, given time to adjust, world supply and demand are relatively responsive (Gardiner and Dixit).

The Role of the United States in World Agricultural Trade

During the 1960s and 1970s world agricultural production grew rapidly enough to increase average world food availability per capita. Even more impressive is the fact, noted above, that during the 1970s world agricultural trade grew more rapidly than agricultural production. The reasons for this rapid growth in agricultural trade are discussed below. The central point is that while world agricultural trade increased from about 225 million metric tons in 1965 to a little over 400 million metric tons in 1980, an increase of 175, the rest of the world's total agricultural export volume grew from about 170 to 225 million metric tons, an increase of only 55. Meanwhile, U.S. agricultural export volume increased 120 million metric tons, significantly more than all the rest of the world combined (see Figure 2) (Figures at the end of chapter).

The U.S. as Exporter. The explosion in U.S. agricultural exports, especially food and feedgrains and oilseeds, is relatively recent, occurring during the late 1960s and 1970s (see Figure 3) (Figures at the end of chapter). The export explosion of the 1970s also appears to have led to expectations about future export demand in the first half of the 1980s that did not materialize. These false expectations are largely responsible for the debt crisis currently facing U.S. agriculture.

One of the best indicators of the importance of the United States in world markets is the U.S. share of major commodities (Figure 4) (Figures at the end of chapter). While market shares for some commodities were quite variable over time, there was a general tendency for U.S. shares to increase rather rapidly during the 1970s, for the U.S. to be a dominant exporter for most of the important traded commodities grown in temperate zones of the world, and for the U.S. share to decline during the 1980s. Even though U.S. market shares have declined during the 1980s, they are still large relative to other major exporters.

Another indication of the importance of the United States as an agricultural exporter is the dependence of major importing countries on the U.S. as a source of supply. During the 1980-84 period, of the 34 largest importers, excluding the United States, 13 obtained more than 20 percent of their agricultural imports from the U.S. and 7 obtained more than one-third of their agricultural imports from the U.S. Among the most dependent on the U.S. in 1984 were Japan, with 36 percent of its imports from the U.S.; Canada, with 39 percent; South Korea, with 48 percent; Mexico, with 81 percent; Taiwan, with 48 percent; and Venezuela, with 47 percent.

The U.S. as Importer. While the importance of the United States as an agricultural exporter is widely recognized, the U.S. role as an importer of agricultural commodities is less well known. During 1980-84, the U.S., on average, was the third largest importer of agricultural commodities with an average total of just over $18 billion per year. The largest was West Germany, with $22 billion, followed by the USSR, with just over $19 billion. In 1980 the United States was the second largest, and in 1984, the world's largest agricultural importer.

Historically, U.S. agricultural imports have consisted primarily of commodities that are not produced in the United States and therefore are not considered directly competitive with domestic production (for example, tropical fruits, coffee, and cocoa). However, as shown in Figure 5 (Figures at the end of chapter), the importation of commodities more immediately competitive with domestic production increased rapidly during the latter part of the 1970s and the first half of the 1980s. During this same period, the value of noncompetitive agricultural imports remained relatively constant until 1980, then declined significantly until resuming growth in 1984. There are several possible explanations for this import behavior. Some of the more plausible include the rapid economic growth of the U.S. economy in the late 1970s and the rapid recovery from recession beginning in 1983. This growth would have contributed to demand for greater variety in the diet of U.S. consumers. A second source of the growth of imports would be the significant increase in the value of the dollar during the 1980s, which reduced the cost of imported commodities to U.S. consumers.

Figure 6 (Figures at the end of chapter) shows the growth in value of imports of groups of commodities that have contributed significantly to both the overall growth of imports and the relative increase in imports of competitive commodities. The U.S. role as an agricultural importer is important to the less developed countries because they are the largest source of agricultural imports into the United States.

U.S. Stockpiling. The third role played by the U.S. in world commodity markets is that of principal stockholder, especially of wheat and coarse grains. The U.S. is by far the largest holder of stocks of these two commodities in absolute terms as well as relative to production. For example, during the 1978/79-1982/83 period, the U.S. accounted for 14.5 percent of world wheat production and 5.2 percent of world wheat use, but 34.5 percent of world wheat stocks (an average of 30 million tons). The other really large holder of wheat stocks was Canada with holdings of 11 million tons, 12.7 percent of the world total. During the same period, the U.S. accounted for 31 percent of world coarse grain production and 21.3 percent of world coarse grain use but held 62.4 million tons or 59.7 percent of world coarse grain stocks. Canada again was the second largest stockholder with 5.6 percent of the total.

This role as the major holder of grain stocks is important to the rest of the world as a stabilizing influence on world markets. The role as the world's shock absorber is not, however, one that the U.S. intentionally chose for itself. Instead, it is a consequence of the domestic commodity price support programs for grains and the fact that support prices have been maintained by buying and holding stocks and, at times, by restricting production (Sharples).

The relatively large stocks held by the U.S. are important to the rest of the world in that they relieve other countries, especially importers, of maintaining their own commodity reserves. The policy of allowing these stocks to be released only at prices significantly higher than support levels means that U.S. stocks do not restrain price except during periods of relatively severe production shortfalls. Thus the combination of relatively high U.S. support prices and stocks has not only provided a sense of security to the world's grain consumers but has also removed a significant amount of downward price risk from foreign producers. This aspect of our stocks and commodity support programs will be discussed more fully below.

The Role of Trade in U.S. Agriculture

As a result of its emergence as the principal exporter of grains and oilseeds during the decade of the 1970s, U.S. agriculture has become more dependent on the rest of the world as a market for its products. One measure of this dependency is the share of U.S. production of major U.S. commodities exported. Figures 7 through 9 (Figures at the end of chapter) present, in order, exports as a percentage of U.S. production for wheat, coarse grains, and soybeans during 1970-85. During this period, the share of exports of production for wheat varied from 37 to 72 percent (Figure 7); for coarse grains, from a low of just over 12 percent in 1970 to a peak of 41 percent in 1983 (Figure 8); and for soybeans and soymeal, from a low of 49 percent in 1974 to a peak of 61 percent in 1981 before declining to 43 percent in 1984 (Figure 9). Figures 10 and 11 (Figures at the end of chapter) present the export share of U.S. production of cotton and rice.

Even though the share of production of crops entering export markets has been quite variable from year to year, there was—with the exception of rice and wheat—a significant upward trend during the 1970s followed by a sharp decline in the 1980s. But even in those years when export share was relatively low, the export market was still very important for these major field crops. Given the slow growth of domestic agricultural demand, any significant withdrawal of American agriculture from world markets would thus entail a shrinkage of the agricultural sector and its supporting agri-industrial infrastructure. Future growth in domestic demand depends essentially on population growth, which has slowed significantly.

Participation in and dependency on foreign markets is important to U.S. agriculture in still another respect. The interdependence of world and domestic markets limits the ability of the U.S. to impose agricultural commodity policies that achieve desired income and equity goals. Most such policies affect the international competitiveness of U.S. commodities; the linkages between world markets and domestic policy are made clearer in the following section.

Changing Trade Environment and Effects on U.S. Exports

The preceding sections of this paper have pointed to the very different performance of U.S. agricultural exports during the latter half of the 1970s in comparison to the first half of the 1980s. This section characterizes differences in the economic and policy environment of these two periods as they relate to the variable U.S. export performance. The variables we focus on include the growth of real income, the international debt of low and middle-income developing countries, the value of the dollar, U.S. farm policy, and the setting of international trade policy.

Both the industrial and developing countries of the world experienced relatively high rates of real economic growth during the 1970s. Growth rates were more variable in the industrial countries as the OPEC oil shock and attempts to deal with it caused growth rates to decline in 1974-75 but were followed by rapid recovery in 1976 (see Figure 12) (Figures at the end of chapter). The developing countries, in contrast, maintained relatively consistent and high rates of real-income growth throughout the 1970s (see Figure 13) (Figures at the end of chapter). The 1980s saw a precipitous decline in the growth of real income in both the industrial and developing countries as efforts to bring inflation under control, led by the United States, produced world recession.

The 1970s was a period not only of rapidly rising aggregate and per-capita income, but also of the rise of international liquidity as the world moved to floating exchange rates, and the large foreign exchange earnings of OPEC were recycled through an increasingly efficient international monetary system. This process made credit available to the food-deficit developing countries at high nominal interest rates but at real interest rates, which, given prevailing rates of inflation, were low and occasionally negative. The combination of rapid income growth, foreign exchange earnings, and the ability to borrow at low real rates created an explosive demand for agricultural commodities by middle-income and oil-exporting developing countries.

The changing monetary and fiscal policies that brought down inflation and slowed economic growth in the early 1980s were accompanied by a significant decline in nominal interest rates but a very significant increase in real interest rates (see Figure 14) (Figures at the end of chapter). This slowing of economic activity, reduced availability of international credit, and significantly higher real interest rates combined with the very rapid increase in debt accumulation by less developed countries to precipitate the widely cited debt crisis of the LDC's. Net borrowing by the LDC's in fact declined rapidly from 1980 through 1983 (see Figure 15) (Figures at the end of chapter). The debt problem of the LDC's of the 1980s thus stemmed not from higher rates of debt accumulation during this period but from the significantly increased debt-service burden, which was a function of declining export revenue and high service obligations on previously accumulated debt (see Figure 16) (Figures at the end of chapter).

The circumstances described above all tended to stimulate general world demand for agricultural imports during the 1970s and to depress such demand in the 1980s. Reinforcing this effect on U.S. agricultural exports was the changing value of the U.S. dollar in world currency markets. The value of the dollar fell significantly against most major currencies throughout the 1970s. This made U.S. agricultural exports cheaper, measured in currencies of importing countries and in currencies of competing exporters, and therefore added to the general increase in demand for agricultural imports. Between 1980 and 1985, the dollar increased in value rapidly (see Figure 17) (Figures at the end of chapter). This caused demand for U.S. agricultural exports to decline more than did the total world demand for agricultural imports.

The domestic farm policy of the United States has been an important factor determining the ability of U.S. agriculture, and thus agricultural exports, to adjust to changing world conditions. For most of the period considered here, U.S. farm policy has consisted of a combination of loan rates, target prices, deficiency payments, government stock programs, and acreage diversion.[1] As world demand for agricultural commodities increased during the 1970s, world prices were above U.S. loan rates so that government stocks did not compete with foreign demand for U.S. production; world prices were transmitted freely to U.S. producers. U.S. farmers responded to strong foreign demand by bringing back into production land that had been idled during the 1960s through land diversion programs and by making other significant investments in increased productive capacity to meet the anticipated growth of demand in the 1980s.

The 1981 farm bill was adopted following a rather long period of high rates of inflation, growing world demand for agricultural commodities, rapidly rising production costs due to spiraling petroleum prices and escalating land values, and the recent experience of export embargoes which imparted a feeling of vulnerability to farmers. The 1981 farm bill, designed to meet the conditions of the late 1970s, contained high and rigid loan rates and target prices. As demand growth slowed in the 1980s and the rising value of the dollar made U.S. commodities relatively expensive in world markets, world prices fell below the mandated loan rates for U.S. products. This made it profitable for farmers to sell their products into government stocks rather than onto export markets. At the same time, foreign producers were provided with continuing incentives to expand their capacity.

Trade policies and practices employed by other countries also influenced U.S. trade performance. The policies of other countries that impact negatively on U.S. agricultural exports fall into two categories: (1) measures that restrict access to markets by U.S. commodities, and (2) subsidies by surplus-producing countries that drive the prices of their exports below domestic price levels. While "unfair" competitive practices have received a great deal of attention as causal factors explaining the decline in U.S. agricultural exports in recent years, these practices have long been followed in essentially their present form by many countries, including the United States. These practices and their consequences have become more visible as world markets have declined and the strong dollar, combined with U.S. commodity price policies, have reduced the cost to our competitors of subsidizing their exports. While trade policies of other countries are important, they probably have not, in and of themselves, been the primary cause of declining U.S. agricultural exports.

Keys to the Future of the United States as an Exporter

The United States during the 1970s enjoyed significant participation in world agricultural markets. Thus far, the 1980s have been characterized by weak markets and unfavorable competitive conditions. The same factors which determined the trade performance of U.S. agriculture during the 1970s and early 1980s are likely to condition our export behavior through the remainder of this decade. We will take a brief look at the apparent direction of these factors, emphasizing the two that are touted as offering the greatest hope--the falling value of the dollar and the 1985 Food Security Act.

The world economy continues to recover from the recent recession but at a rate much slower than had been expected. The international financial system seems to have found a way to cope with the debt problems of the less-developed countries, but it appears that debt service requirements will dampen agricultural import demand by LDC's for at least several more years. A potentially more serious consequence of the debt of these countries is the effect that austerity programs and structural adjustment may have on their long-term growth and hence on their long-term demand for agricultural imports. Nor does it appear that we will have any one-time boost in world agricultural demand similar to that provided by the entry of the Soviet Union as a significant importer or by the opening of the People's Republic of China to world markets. Aggregate world demand for agricultural imports will therefore grow through the remainder of this decade but probably at a rate significantly lower than that of the later 1970s.

U.S. agricultural policies and the high value of the dollar also served as a protective price umbrella over foreign producers. Many foreign countries responded by significantly increasing productive capacity. Much of this capacity is in the form of fixed assets, so that downward adjustment in foreign production in response to lower world prices will likely be very gradual. The United States will thus have to compete with a larger foreign supply than it would have in the absence of this capacity.

The trade policy environment contains both positive and negative tendencies. On the positive side, planning is under way and a schedule has been established for the opening of a new round of multilateral trade negotiations under the General Agreements on Tariffs and Trade. There seems to be strong support, as evidenced by the communique from the 1986 Tokyo Summit, for making agriculture a central focus of this round of negotiations. Also positive is some progress in resolving disputes between the United States and Japan and discussions relating to establishment of a U.S.-Canadian free trade zone. On the negative side, there has been an increasing level of protectionist sentiment in the U.S. and other countries. The world presently stands at a watershed from which it could move to significant improvement in international trade or slide into a very serious and damaging protectionist war.

A force that has created considerable optimism for a turnabout in U.S. agricultural exports is the decline in the dollar. Since the dollar peaked against the Japanese yen and major European currencies in early 1985, it has fallen against this basket of currencies by approximately 30 percent (see Figure 18) (Figures at the end of chapter). However, there are two factors which should temper expectations for an induced rapid increase in agricultural exports. First, the fall in the dollar has been selective--not uniform against all currencies. Figures 19 through 23 (Figures at the end of chapter) show the value of the dollar over the 1970-85 period weighted against different currency groupings. It can be seen that from the early 1980s until the beginning of 1985, the dollar rose relatively uniformly against the weighted average of currencies of the major importers of U.S. wheat, corn, and soybeans, as well as against the currencies of our major competitors in these commodities. However, the decline in the dollar has been relatively minor against the currencies of major importers of wheat and against our major competitors in all three commodities. Thus, we cannot expect large gains in exports of these major bulk commodities due to the depreciation of the dollar that has occurred thus far.

The second factor that should be considered in shaping expectations as to the impact of the exchange rate is that a considerable time lag is required for a decline in the dollar to work its way through the financial and commodity markets and exert its full impact on exports. Research conducted at the Economic Research Service (Krissoff and Morey) indicates that approximately 10 quarters are required for the full effect to be felt. On the positive side, the same research does indicate that exports are responsive in the longer term to the relative value of the dollar.

The Food Security Act of 1985, more commonly known as the 1985 Farm Bill, evolved in a very different environment than did the 1981 legislation.[2] The 1985 bill was an effort to shape a market-oriented policy that would eliminate features of the earlier legislation that reduced competitiveness of U.S. exports. Three provisions of the legislation should help to remove government programs as a wedge between world markets and the U.S. farmer and allow U.S. agriculture to respond to changing world conditions. First, loan rates are made flexible. Base loan rates are calculated by a moving average of world market prices, and the Secretary has authority to lower loan rates even further if he deems this necessary to enhance U.S. competitiveness. This feature of the legislation should allow our goods to be price competitive while removing the price umbrella from foreign producers. A second feature fixes program yields and therefore removes the incentive for farmers to use variable inputs more intensively in order to increase yield and therefore increase deficiency payments. A third feature breaks the link between the size of deficiency payment for which the farmer is eligible and the number of acres of the crop planted. The farmer will receive full deficiency payments if he plants more than 50 percent of his base. This feature is intended to let the farmer respond to market price signals rather than to program signals in making planting decisions.

On the negative side, target prices remain relatively high and may still provide farmers with the incentive to produce more than market prices justify. These high target prices and associated deficiency payments can also be perceived by the rest of the world as implicit U.S. export subsidies. Another negative factor is the failure of the legislation to significantly liberalize our sugar and dairy programs, which are protectionist and are used against us in trade negotiations with our competitors. The legislation also incorporates a number of features that provide, and in some cases mandate, retaliation against competitors for policies and practices deemed to be unfair. These additional trade policy weapons may be useful as bargaining chips in future negotiations. But they could fan the flames of a smoldering, potential trade war, depending upon how they are used.

In summary, it seems reasonable to expect that U.S. export performance will be better than that achieved during the first half of the 1980s but we are not likely, in the foreseeable future, to experience the kind of export boom that occurred during the 1970s. What does remain clear is that the international market environment and the policy environment will be critical in determining the well-being of the U.S. agricultural sector and the farm population.

Notes

1. The reader desiring a brief, concise description of U.S. agricultural policy is referred to Webb. A brief definition of terms is presented here.

Loan Rate--
The price per unit at which the Government will provide loans to farmers to enable them to hold their crops for later sale. Commodity may be forfeited to the Government as repayment in full of the loan if market price is below the loan rate.

Target Price--
A price level established by law for specified crops which is used in the determination of deficiency payments.

Deficiency payment--
Government payment made to farmers who participate in specified programs; payment rate is based on the difference between a target price and the market price or the loan rate, whichever difference is less.

Government stocks payment--
Stocks of specified commodities held by farmers (Farmer Owned Reserve) or the Commodity Credit Corporation (CCC) as collateral for commodity loans or stocks owned by CCC which have been forfeited by farmers.

Acreage diversion--
A variety of programs which require or provide incentives for farmers to divert land from production of specified crops (usually to a soil conserving use).

2. For a complete description of the provisions of the Food Security Act of 1985 see Glaser.

References

Gardiner, Walter H. and Praveen M. Dixit, Price Elasticity of Export Demand--Concepts and Estimates, Economic Research Service, U.S. Department of Agriculture, Staff Report No. AGES860408, May 1986.

Glaser, Lewrene K., Provisions of the Food Security Act of 1985, Economic Research Service, U.S. Department of Agriculture, Information Bulletin No. 498, April 1986.

Krissoff, Barry and Art Morey, The Dollar Turnaround and U.S. Agricultural Exports, Economic Research Service, U.S. Department of Agriculture, Staff Report No. AGES861128, Dec. 1986.

McCalla, Alex F. and Timothy E. Josling, Agricultural Policies and World Markets, MacMillan Publishing Company, New York, 1985.

Sharples, Jerry A., Is the United States the World's Residual Supplier of Grain?, Staff Paper No. 85-9, Department of Agricultural Economics, Purdue University, July 1985.

Webb, Alan, "The Evolution of U.S. Agriculture and U.S. Agricultural Policy," Chapter 3 in Embargoes, Surplus Disposal, and U.S. Agriculture, Economic Research Service, U.S. Department of Agriculture, AER No. 564, Dec. 1986.

Table 1--Variability in Agricultural Production, Selected Countries/Regions 1/

Item	1961-72	1972-83
	Percent	
Production Variability		
United States	1.6	3.5
EC-10	2.1	3.2
Australia	4.1	5.5
USSR	5.0	6.1
Middle America	1.1	3.0
North Africa/Middle East	2.9	3.9
East Asia	4.4	7.1
World	1.5	1.6

1/ Measured as the coefficient of variation from best-fit linear or curvilinear time trends.

Source: Economic Research Service, USDA.

Table 2--Percent of World Corn and Wheat Imports Handled by Various Marketing Arrangements, 1960-80

	1960	1965	1970	1975	1980
Corn					
Free traders	15.9	13.4	22.1	19.2	25.0
State traders	7.5	6.6	12.0	40.6	47.7
Variable levies	76.6	80.0	65.9	40.1	27.3
Wheat					
Free traders	2.9	2.7	5.2	4.3	3.2
State traders	62.6	77.9	65.4	75.1	80.9
Variable levies	34.5	19.4	29.4	20.6	13.6

Source: Economic Research Service, USDA.

Figure 1 -- Ratio of World Trade to World Production in Wheat and Coarse Grains, 1960-85

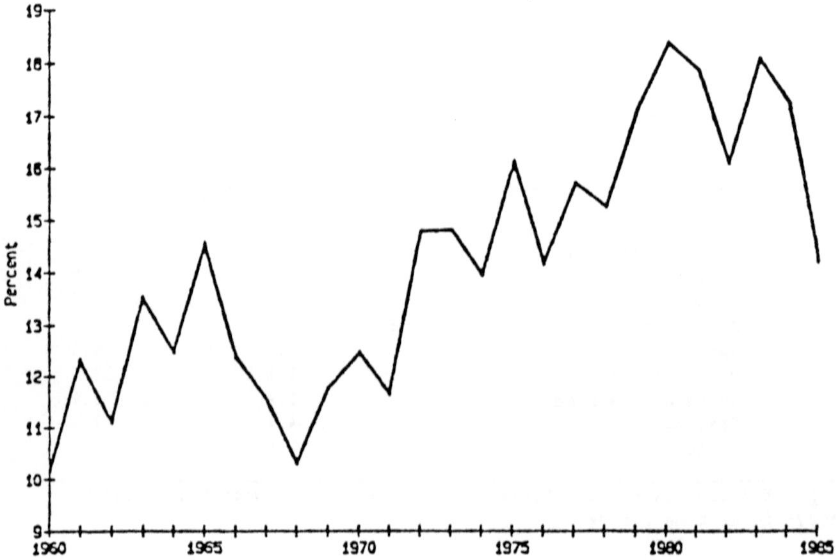

Source: USDA, FAS.

Figure 2 -- Agricultural Export Volume -- U.S. and Rest of World 1965-85

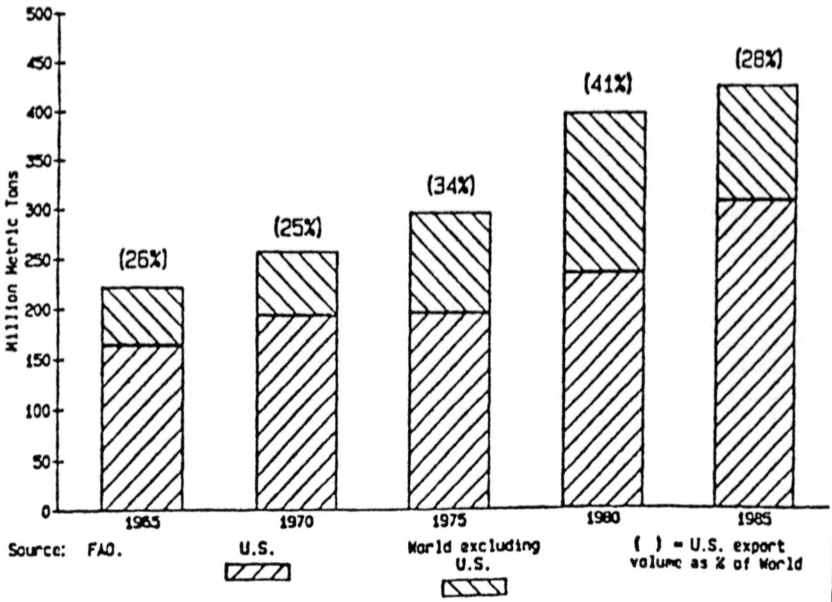

Source: FAO.

Figure 3 -- Value of U.S. Farm Exports, 1935-85

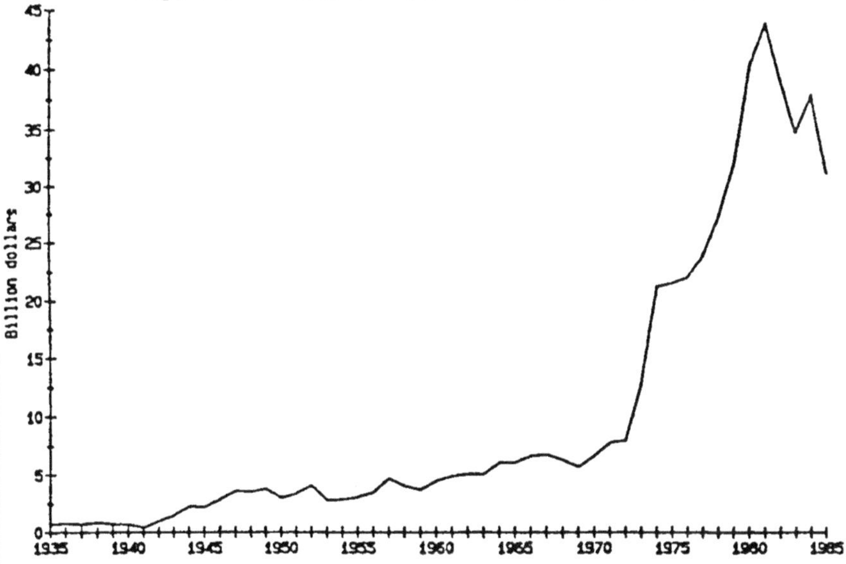

Source: USDA, ERS.

Figure 4 -- U.S. Share of World Trade in Major Commodities 1970 - 1985 (% of World Volume)

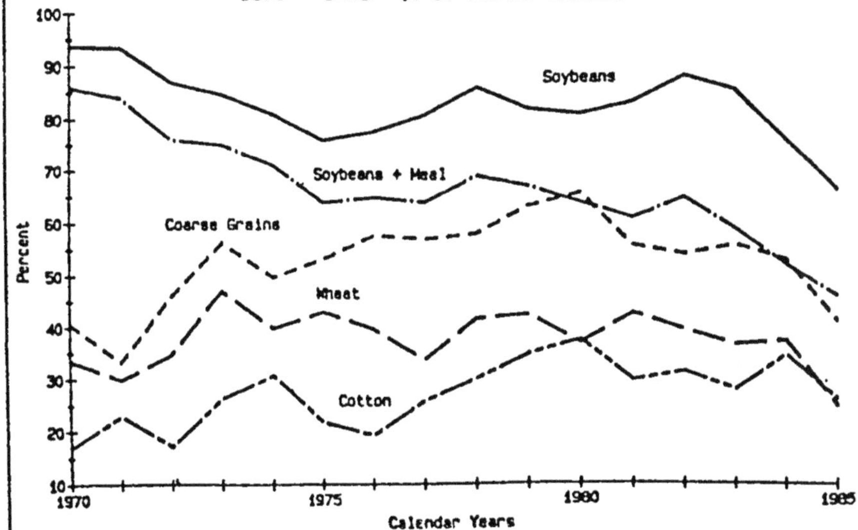

Source: FAO.

Figure 5 — Value of Competitive and Noncompetitive U.S. Agricultural Imports, 1970-85

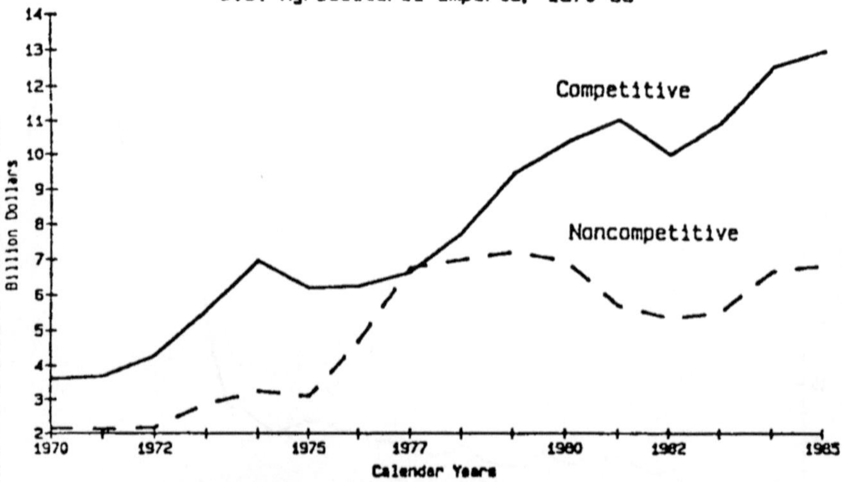

Source: USDA, ERS. Note: Competitive imports are those for which domestic U.S. production of the same or a close substitute product is significant while noncompetitive imports have no close substitute produced in significant quantities domestically.

Figure 6 -- Value of Selected U.S. Agricultural Imports 1970-85

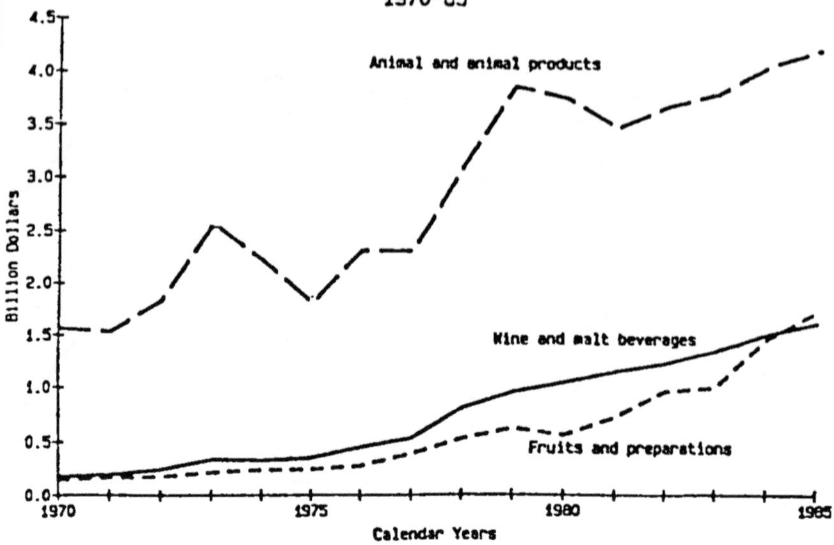

Source: USDA, ERS.

Figure 7 — U.S. Wheat Exports
As Share of U.S. Production, 1970-85

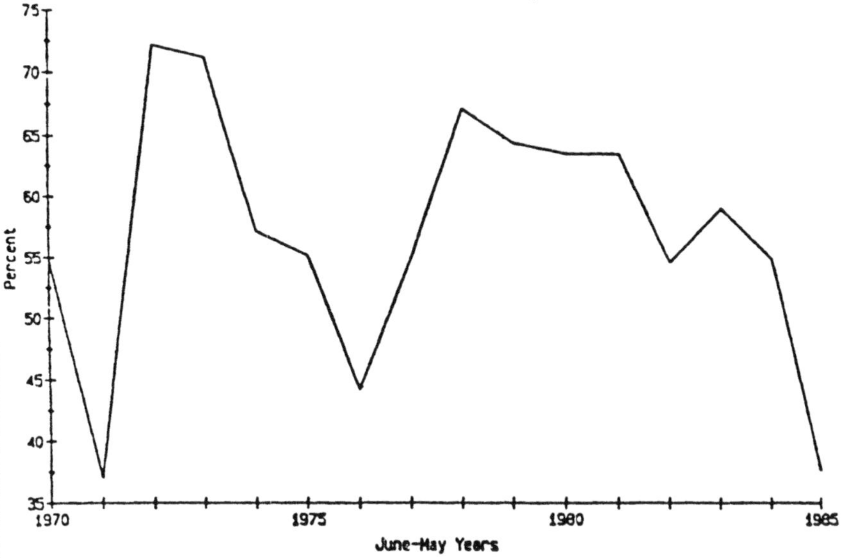

Source: USDA, ERS.

Figure 8 — U.S. Coarse Grain Exports
As Share of U.S. Production, 1970-85

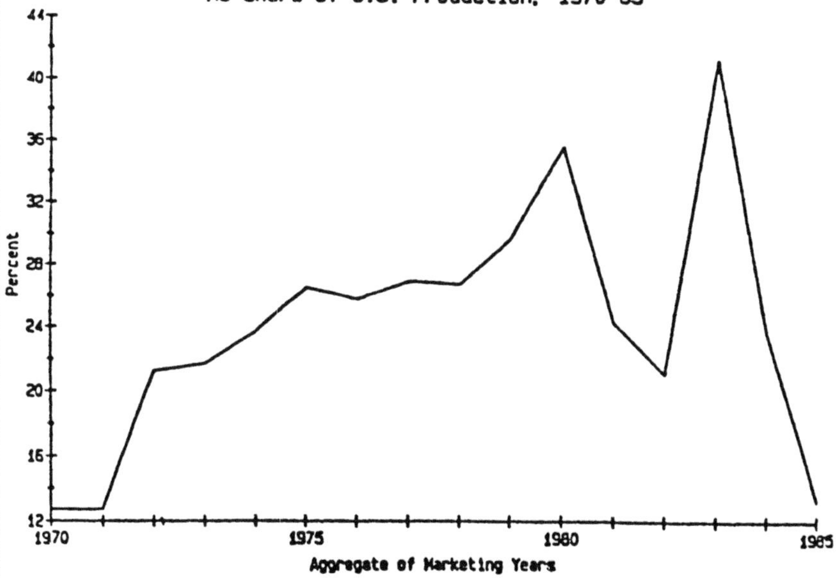

Source: USDA, ERS.

Figure 9 -- U.S. Soybean Exports as a Share of U.S. Production 1970-85

Source: USDA, ERS.

includes meal in bean equivalents
Sept-Aug Years

Figure 10 — U.S. Cotton Exports As Share of U.S. Production, 1970-85

Source: USDA, ERS.

Figure 11 -- U.S. Rice Exports
As Share of U.S. Production, 1970-85

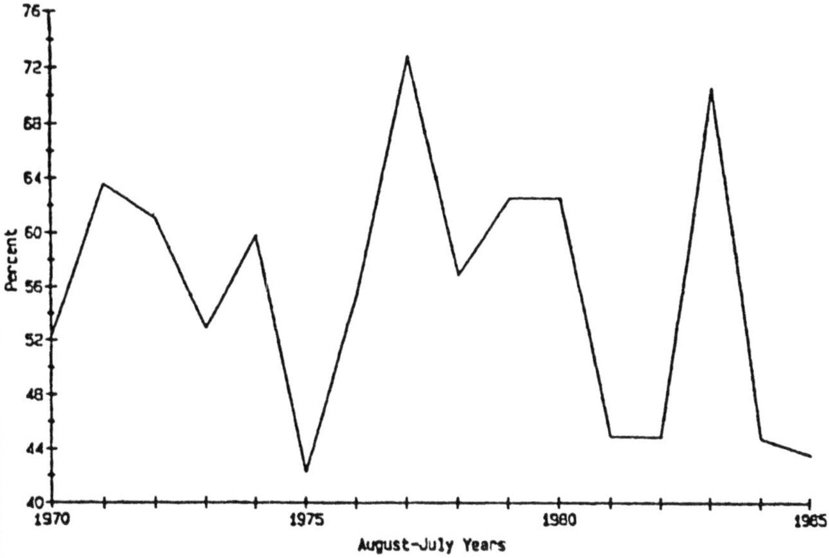

Source: USDA, ERS.

Figure 12 -- Real Growth in Income
Developing Countries

Source: International Monetary Fund.

Figure 13 -- Real Growth in Income
Industrial Countries

Source: International Monetary Fund.

Figure 14 -- Nominal/Real Interest Rates
U.S. Money Market Rate

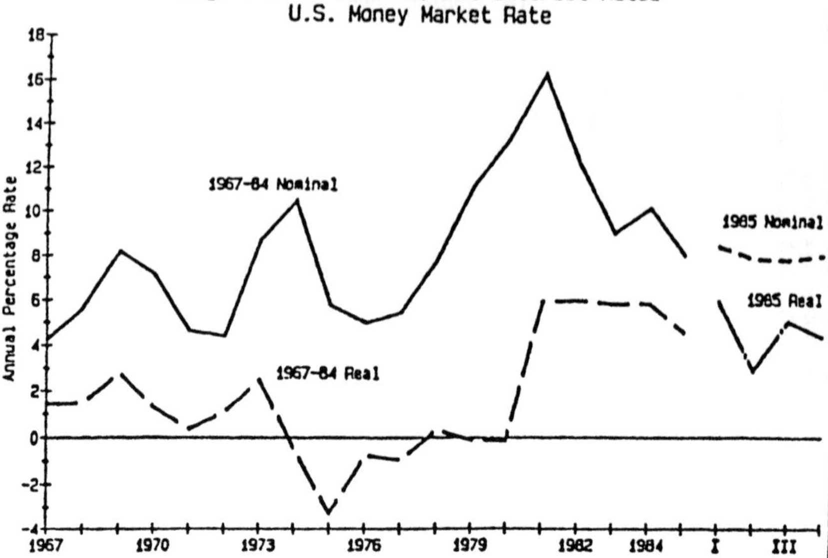

Source: IMF.

Note: 1967-84 rates are annual averages while 1985 rates are quarterly averages.

Figure 15 -- Total Debt, All LDC's. 1978-85
Percentage Changes

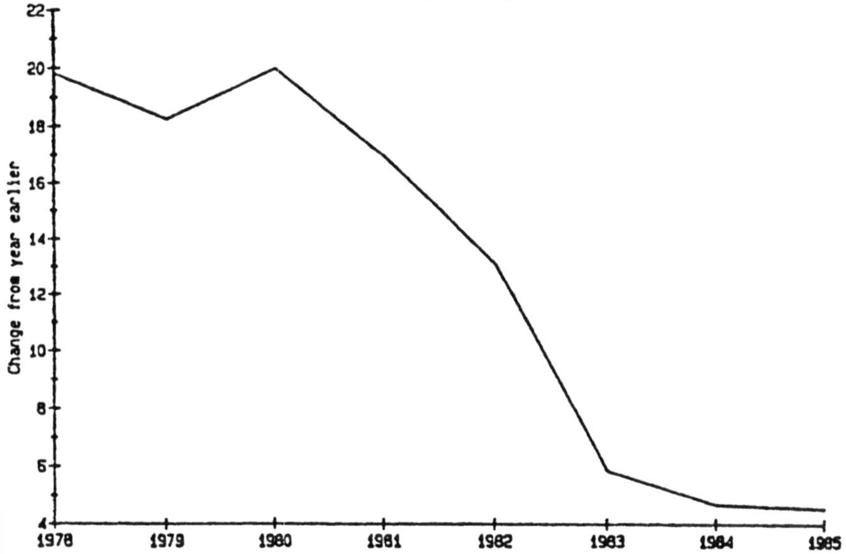

Source: International Monetary Fund.

Figure 16 -- Debt Service Ratio, 1977-85
All LDC's and Latin America

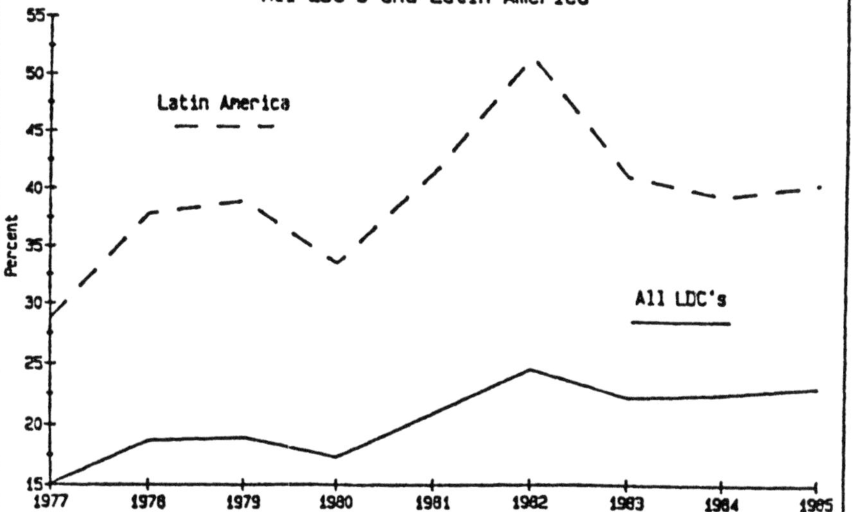

Source: International Monetary Fund.

Note: Debt service ratio is the sum of principal and interest due on foreign debt as a percentage of foreign exchange earnings.

Figure 17 -- Index of Weighted -- Average Dollar Exchange Rate, 1970-85
Federal Reserve Formula

Source: IMF.

Note: Values for 1970-84 are annual averages while values for 1985 are quarterly averages.

Figure 18 -- Index of Real Weighted -- Average Dollar Exchange Against Currencies of Major Importers of U.S. Wheat, 1970-85

Source: IMF.

Note: Values for 1970-84 are annual averages while values for 1985 are quarterly averages.

Figure 19 -- Index of Real Weighted -- Average Dollar Exchange Against Currencies of Major Importers of U.S. Corn, 1970-85

Source: IMF.

Note: Values for 1970-84 are annual averages while values for 1985 are quarterly averages.

Figure 20 -- Index of Real Weighted — Average Dollar Exchange Against Currencies of Major Importers of U.S. Soybeans, 1970-85

Source: IMF.

Note: Values for 1970-84 are annual averages while values for 1985 are quarterly averages.

Figure 21 -- Index of Real Weighted -- Average Dollar Exchange Against Currencies of Major Competitors of U.S. Wheat Exports, 1970-1985

Source: IMF.

Note: Values for 1970-84 are annual averages while values for 1985 are quarterly averages.

Figure 22 -- Index of Real Weighted -- Average Dollar Exchange Against Currencies of Major Competitors of U.S. Corn Exports, 1970-1985

Source: IMF.

Note: Values for 1970-84 are annual averages while values for 1985 are quarterly averages.

Figure 23 -- Index of Real Weighted -- Average Dollar Exchange Against Currencies of Major Competitors of U.S. Soybean Exports, 1970-1985

Source: IMF.

Note: Values for 1970-84 are annual averages while values for 1985 are quarterly averages.

3

World Food Situation in Historical Perspective

John W. Mellor

Introduction

The current world food situation is dramatically different from what it was a decade ago. In the mid-1970s, the world seemed beset by major food shortages; today, however, it appears to be awash in food. In the developed countries of Europe, Australia and the United States food surpluses are currently piling up at record rates. Ten years ago, the food problems of the world seemed to focus on Asia; today, however, such concerns focus mainly on Africa. In the mid-1980s, Africa is the only region of the world still plagued by major food production shortfalls.

Under these changed circumstances, it might be tempting to suggest food trade as the 1980's solution to the world food problem. Shipping food from "surplus" countries--like Europe and the United States--to "deficit" countries--like Africa--might seem to represent the easiest solution to this decade's world food problems.

Such a simple-minded solution would, however, neglect two very important factors. Firstly, hundreds of millions of poor people at virtually any price. These poor people suffer from the disabilities caused by a lack of income and employment opportunities. Thus, increased food trade between the developed and the developing countries of the world must be coupled with efforts to raise the purchasing power of low-income people throughout the world.

Secondly, in most developing countries, accelerated agricultural growth represents the best means for stimulating overall economic growth. In most cases, agriculture can play a number of pivotal roles in the development process. Increased agricultural production boosts domestic food supplies at the same time that it stimulates further rounds of employment growth in the service and urban sectors of the economy. Moreover, because of its linkage effects with the rest of the economy, agricultural growth helps raise the poor's access to food supplies.

Successful pursuit of such an agricultural strategy of development requires much from the developed countries of the world. In particular, it requires the supply of food imports that--rather surprisingly--accompany almost any strategy of economic growth in the Third World. During the last decade food imports to the developing world have been rising steadily, as rapidly increasing growth in the demand for food continues to exceed local production capabilities. This tendency seems likely to continue well into the 1990s. Such forces should make it even more attractive for Third World countries to focus their attention on the demands of accelerated agricultural development.

Past Trends in Population, Production, and Consumption

In order to understand the need for an agricultural strategy of development in the Third World, it is necessary to examine past trends in food production and consumption.

Between 1966 and 1980 food production in the world as a whole increased at an average rate of 2.8 percent a year (Table 1) (Tables at the end of chapter). This was slightly faster than the average annual population growth rate of 2.0 percent. On a per capita basis, world food production increased by 0.8 percent.

However, this aggregate figure covers sharply different rates of food production growth in various regions of the world. For example, in the developed countries, per capita food production increased by an impressive 1.5 percent between 1966 and 1980. This was three times the annual rate of per capita food production growth (0.5 percent) recorded in the developing countries.

Rates of food production growth have also varied considerably within the developing world. For instance, in an area of the Third World that was once considered quite famine prone (Asia), per capita food production increased by a strong 1.0 percent. However, in the new food deficit area of the world (Sub-Saharan Africa), per capita food production fell by a shocking 1.4 percent.

The sharp decline in food production in Sub-Saharan Africa is so striking as to command our immediate attention. On the one hand, the roots of this food crisis in Africa go back a long way, to include such factors as a series of poor crop years, low-historic levels of government investment in agriculture and unfavorable public policies vis-a-vis agriculture. Yet at the same time, the crisis also includes the notable absence of proven technological packages for small farmers in most of the rainfed farming systems of Africa. To date, the new seed-fertilizer technologies commonly associated with the Green Revolution have not had much of an aggregate impact on Africa (Eicher, 1982).

According to Table 1, in recent years, the growth rate of food production in the developed world has exceeded the rate of food consumption growth. This is primarily a result of the relatively low rates of food consumption growth in the industrialized countries of Western Europe, Australia, and the United States. However, in the developing countries of Africa, Asia, and Latin America, the situation is quite different. In these countries, rapidly increasing rates of food demand growth have generally outstripped the productive capabilities of domestic agriculture. As a result, net imports of cereals to the developing world (excluding China) rose from 14 million metric tons in the late 1960s to nearly 50 million tons in the early 1980s. I will have more to say about the importance of these food imports shortly.

Variability and Food Production in the Third World

In analyzing past trends in food production growth in the Third World, it is important to realize that such growth has been accompanied by a steadily increasing degree of production variability. According to research at the International Food Policy Research Institute (IFPRI) (Hazell, 1984), between the periods 1960/61 to 1970/71 and 1971/72, the coefficient of variation of total world cereal production increased from 2.8 percent to 3.4 percent. This represented a net increase in production variability of 21 percent (Table 2) (Tables at the end of chapter).

Preliminary analysis suggests that the major source of this increase in production variability lies in increases in yield covariances between crops and regions.

This may well be because of factors associated with the new seed/fertilizer technologies. For example, if all of a country's production of a crop—such as maize in the United States—has a single parent, that crop might be more vulnerable to pestilence. This is a problem that crop scientists are currently analyzing. Yet in the meantime, another problem still remains. In many developing countries, policies affecting the availability of fertilizer, electricity, and water inputs change from year to year. Such policy changes may have a large and unfavorable effect on agricultural production as that production becomes more dependent on the supply of those water and fertilizer inputs that are associated with the new technology.

In recent years, the steady growth in world food production has also been accompanied by a rising degree of price variability. While international grain prices were relatively stable in the 1950s and 1960s, since 1971 they have become highly variable. According to IFPRI research (Valdes, 1984), the for wheat coefficient of variation for export prices is almost eight times higher in the 1970s than it was in the 1960s (Table 3) (Tables at the end of chapter). For rice, the coefficient of variation for export prices more than doubled between the two decades.

As noted above, in recent years, food imports to the Third World have been steadily increasing. The reasons for this are as follows. During the development process, two principal forces tend to fuel a steady rise in the demand for food: population growth and per capita income growth. The manner in which these two dynamic forces interact is illustrated in Table 4 (Tables at the end of chapter), which depicts five stylized phases of food demand and economic growth.

Row one of the table shows an early stage of economic growth in which people are very poor, desperately wishing to consume more food, yet unable to do so because of low incomes. In this stage, poverty causes high death rates and hence, only modest rates of population growth. The result is a 3 percent or less growth rate in the effective demand for food—a rate that can be met by more effort on a slightly expanded land base.

As development occurs, the population growth rate increases. But, even more importantly, income begins to grow rapidly and the two together increase the growth rate of demand for food by some 30 percent over the earlier phase. Such a rate of growth in food demand exceeds all but the most rapid rates of food production growth. Thus a high rate of technological change in agriculture is needed in this stage of development (Row 3 of the table).

However, in recent years, even those countries with the most impressive rates of technological change in agriculture have been unable to meet their rates of food demand growth. For example, the 16 developing countries with the fastest growth rates in basic food staples production over the period 1961-76 collectively more than doubled their net food imports (in tons) during this time period (Bachman and Paulino, 1979). These data demonstrate that most countries in the high growth, medium income stage of development find it necessary to rely upon food imports to meet a portion of their surging food demand growth.

In the later stages of development, of course, population growth rates decline and growth in income begins to have little effect on the demand for food. Meeting food demand growth then becomes more manageable, particularly since by then food production growth rates have become institutionalized at high levels. It is in this stage that food imports become unnecessary and agricultural surpluses begin to accrue.

In the modern Third World, many developing countries are currently in the high growth, medium income stage of development. They are, therefore, increasingly dependent on food imports to meet their domestic food needs. According to Table 5 (Tables at the end of chapter), between the years 1966-70 and 1976-80, net food imports to the Third World increased by more than threefold, from 12 to 38 million metric tons per year. Since then, they have increased even more.

A close reading of the data in Table 5 suggests that increasing per capita income is the dynamic factor underlying the surge in food imports in the Third World. For example, referring to the table, countries with the highest rates of per capita income growth experienced a 10.1 percent annual increase in food imports. The countries in the next highest growth category also had a high rate of increase in food imports, as they more than doubled their level of imports during the study period. The only exception to this finding is the relatively slow growth countries (between 1.0 and 2.9 percent GNP per capita increase).

While increasing per capita income fuels food demand growth in the Third World, it is important to realize that, as income rises, the relative character of that food demand changes. Rising demand causes food demand to shift to the more preferred cereals, and to highly income-elastic livestock products. The latter, in particular, become increasingly important in the consumption patterns of consumers, as evidenced by the fact between 1961-65 and 1973-77 annual meat consumption in the Third World increased at a 3.4 percent rate, significantly faster than population growth (Sarma and Yeung, 1985).

The rising importance of livestock products in developing countries plays a major role in restraining the decline in the overall income elasticities for basic food staples. The income elasticity of demand for livestock products remains fairly stable up to relatively high-income levels. This results in a strong derived demand for basic food staples, even as income rises. This phenomenon is reinforced by the fact that among livestock products demand for pigs and poultry, both of which are produced at the margin largely on concentrate feed, grows most rapidly.

It is instructive to note here a peculiarity of the derived demand for feed for livestock and its effect on the aggregate income elasticity of demand for basic food staples. At low incomes, livestock commodities comprise a small budget share, and hence the derived demand for basic food staples is quite small. As incomes rise, the income elasticity of demand for basic food staples for direct human consumption declines; but at the same time, the income elasticity of demand for food staples for livestock consumption begins to increase. Initially, the base level of the derived demand is very small relative to direct demand, but, with sharply different elasticities, the relative weights change quite rapidly. Thus the income elasticity of demand for total food staples forms an S-shaped curve, with the weighted average elasticity first declining, then rising, and then eventually declining again. It is in the period when the weighted average elasticity of both direct and derived demand peaks that developing countries move onto the international market for substantial aggregate imports of food. Given that imports are initially small relative to total consumption and hence highly leveraged, it becomes clear why analysts are normally caught unawares by the explosive growth in imports of basic food staples by developing countries after a period of declining or slow growth in imports.

Towards an Agricultural Strategy of Development

The surging rate of food demand growth in the Third

World must be met largely through technological change in agriculture. Technological inputs--such as high-yield seeds, fertilizers, and irrigation systems--play a critical role in virtually all modern methods of agriculture. For example, the adoption of new agricultural technology in India helped increase cereal yields 29 percent between the periods 1954/55 to 1964/65 and 1967/68 to 1977/78 (Alagh and Sharma, 1980). Agriculture is particularly dependent on improved technology for growth because of the limited capacity to expand land areas.

Technological inputs stimulate agricultural output by raising crop yields. Throughout the world, even in Africa, the rate of growth of the cropped area has declined sharply in recent years. Thus an ever-increasing proportion of the food needed to feed the world must come from increased yields per unit of land.

In the past two decades, crop yields have, in fact, become the main source of food production growth in the developing world. Between 1961 and 1980, output per hectare of major food crops in the developing world rose by 1.9 percent annually, and accounted for more than 70 percent of total food production growth (Table 6) (Tables at the end of chapter). During this period, increases in the harvested area averaged only 0.7 percent a year, and contributed the other 30 percent of total production growth in the Third World.

In addition to boosting crop yields and improving overall agricultural production, technological change in agriculture can also help stimulate broader patterns of rural and economic change. In most situations, technological change in agriculture can play three central roles in the overall development process.

Firstly, it is important to recognize that food and employment represent two sides of the same coin. In the developing world, low-income people typically spend the bulk of additional income on food. In most developing countries, average budget shares for food among the poor range between 60 and 80 percent. Thus any strategy of development that leads to a rapid growth in the employment and income of the poor, also leads to a large increase in the effective demand for food. If more food is not forthcoming, food prices will rise, the real cost of labor will increase, and investment will swing to more capital-intensive processes (Mellor, 1976). Thus any strategy of development that entails more employment for the poor will also require the wage goods--particularly food--to support such economic growth. In this sense, a high employment policy is also a high food demand policy.

Second, technological change in agriculture has important employment and income linkages with the rural nonfarm economy. Technological change in agriculture raises the incomes of landowning farmers, who spend a large proportion of their new income on a wide range of nonagricultural goods and services. In Asia, for example, farmers typically spend 40 percent of their increments to income on such locally-produced, nonagricultural goods and services as textile products, transportation and health services, and housing (Hazell and Roell, 1983). The small enterprises that produce such goods tend to be far more labor-intensive than any fertilizer factory or steel mill. These enterprises thus provide the rural poor with a whole spectrum of new nonagricultural employment opportunities. This increases the effective purchasing power of the poor at the same time that it provides for new rounds of growth in the economy at large. As the poor begin to work regularly, they demand more and higher-valued foodstuffs. This helps to stimulate the demand for foodstuffs and to strengthen the need for further technological change in agriculture.

Third, technological change in agriculture encourages employment growth in the urban sectors of the economy. Inexpensive food helps keep labor costs down, thereby increasing the comparative advantage inherent in labor-intensive exports. In addition, technological change in agriculture stimulates domestic demand for those labor-intensive, consumer goods--such as clothing and textiles--in which developing countries possess a distinct comparative advantage. Over time, firms specializing in the production of these commodities can acquire the experience and efficiency needed to compete on the world market. This is important, inasmuch as any successful strategy of development requires the production of export goods to pay for a wide range of capital-intensive goods--for example, fertilizer and pesticides for agriculture, and steel and petrochemicals for industry. A strategy of technological change in agriculture, which stresses the increased production of primary and consumer goods, is able to contribute to these export needs.

In the early stages of development, technological change in agriculture helps produce the agricultural commodities that are needed to earn foreign exchange. In the latter stages, technological change in agriculture helps create the domestic demand needed to facilitate the growth of those labor-intensive industries that can compete on the world market. Taiwan is a good case in point of a country which used an agricultural strategy of development to create small-scale manufacturing and industrial enterprises that could compete on the world market.

The relationship between technological change in agriculture and employment growth is highly complementary and must be a major focus of policy. In Asia, with the Green Revolution underway, the focus needs to be on seeing that capital allocations are efficient so as to keep employment growth commensurate with the improved agricultural record. Several Asian countries are now deficient in this respect. In Africa, however, the effort needs to be more on instituting technological change in agriculture, simply to get the now-stagnant rural sector moving. While per capita food production has been increasing in recent years in the developing world as a whole, in Africa it has been falling. During the period 1966-80, per capita food production in Sub-Saharan Africa fell an alarming 1.4 percent per annum (Table 1). Clearly, much needs to be done in Africa in order to get the food production sector moving.

Conclusion

During the next few decades, development in the world food situation will present new challenges and responsibilities for three sets of actors: the developing world, the developed world, and the research community.

In the developing world, many countries need to pay more attention to agriculture. At present, the low purchasing power of the poor in many Third World countries means that they lack the means to buy more food at <u>any</u> price. In most of these situations, technological change in agriculture represents the best way to boost the purchasing power of the poor. Technological change in agriculture raises the total level of domestic food supplies at the same time that it increases the ability of the poor to buy such supplies.

Accelerated growth in agriculture, however, requires that more developing countries pay more attention to the financial requisites of rural growth. The leaders of these countries need to raise their allocations to the basic building blocks of agricultural development: irrigation, agricultural research, and rural roads. They must also revise their pricing and exchange rate policies vis-a-vis the agricultural sector.

The developed world, in turn, must seek to encourage such policy reappraisals by making available the technical and financial resources needed to support an agricultural strategy of development in the Third World. They must also be prepared to make available the food imports (and food aid) that accompany the process of agricultural growth in the Third World.

The research community must support these larger processes by focusing their energies on current problem areas in the world. For example, in dealing with the present food dilemma in Africa, researchers need to be much more concerned with effective demand for food. Some of this effective demand will come from increased incomes and employment in agriculture, but a substantial portion will have to come from accelerated employment growth in the nonagricultural sector. We clearly need to know much more about the sources of such nonagricultural employment growth, and its effects on the marginal propensity of the poor to spend wage income on different kinds of food commodities.

From the dynamics of such a three-way partnership, the world could conceivably evolve into a place where adequate food is not just a right of all people, but an accepted fact.

REFERENCES

Alagh, Y.K. and Sharma, P.S. 1980. "Growth of Crop Production: 1960/61 to 1978/79—Is it Decelerating?" Indian Journal of Agricultural Economics, Vol. 35, No. 3 (April-June).

Bachman, Kenneth L. and Paulino, Leonardo A. 1979. Rapid Food Production Growth in Selected Developing Countries. Research Report 11, International Food Policy Research Institute: Washington, DC.

Eicher, Carl. 1982. "Facing Up to Africa's Food Crisis." Foreign Affairs, Vol. 62, No. 1 (Fall).

Hazell, Peter. 1984. "Sources of Increased Variability in World Cereal Production Since the 1960s." Journal of Agricultural Economics, Vol. XXXVI, No. 2 (May).

Hazell, P. and Rell, A. 1983. Rural Growth Linkages: Household Expenditure Patterns in Malaysia and Nigeria. Research Report 41, International Food Policy Research Institute: Washington, DC.

Lele, Uma. 1981. "Rural Africa: Modernization, Equity, and Long-Term Development." Science, Vol. 211 (February 6, 1981).

Mellor, John. 1966. The Economics of Agricultural Development. Ithaca: Cornell University Press.

_________. 1978. "Food Price Policy and Income Distribution in Low-Income Countries." Economic Development and Cultural Change, Vol. 27, No. 1 (October).

_________. 1985. "Opportunities in the International Economy for Meeting the Food Requirements of the Developing Countries of the World." Paper presented at Conference on the Political Economy of Food, Utah State University, Logan, Utah. May 1985.

Mellor, John and Ranade, Chandra. Forthcoming. "Technological Change in a Low-Labor Productivity, Land Surplus Economy: The African Development Problem."

Paulino, Leonardo A. Forthcoming. "Food in the Third World: Past Trends and Projections to 2000." International Food Policy Research Institute: Washington, DC.

Sarma, J.S. and Yeung, Patrick. 1985. Livestock Products in the Third World: Past Trends and Projections to 1990 and 2000. Research Report 49, International Food Policy Research Institute, Washington, DC.

Valdes, Alberto. 1984. "A Note on Variability in International Grain Prices." Unpublished paper prepared for IFPRI Workshop on Food and Agricultural Price Policy. Washington, DC: International Food Policy Research Institute.

Table 1--Growth rates of population, production, and consumption of major food crops[a] in developing and developed countries, 1966-1980

Country Group	Ave. annual Population Growth Rate, 1966-1980	Ave. annual Production Growth Rate, 1966-1980	Ave. annual Consumption Growth Rate, 1966-1980
	(percent)	(percent)	(percent)
World	2.0	2.8	2.6
Developed countries	1.0	2.5	1.9
By region			
EEC	0.7	2.1	1.4
Eastern Europe and U.S.S.R.	0.9	2.4	2.8
United States	1.2	3.1	0.8
Others	1.3	2.0	2.0
Developing Countries	2.4	2.9	3.3
By region			
Asia[b]	2.3	3.3	3.3
North Africa and Middle East	2.7	2.6	3.9
Sub-Saharan Africa	3.0	1.6	2.3
Latin America	2.5	2.2	3.1

Source: Paulino (forthcoming).

[a] Includes cereals, roots and tubers, pulses, groundnuts, bananas, and plantains. Rice is in husked form at 80 percent of paddy.

[b] Includes the People's Republic of China.

Table 2--Changes in the mean and variability of world cereal production;[a] 1960/61-1970/71 to 1971/72-1982/83

	Average Production			Coefficient of Variation of Production		
	First Period	Second Period	Change	First Period	Second Period	Change
	..... (Metric tons)			.... (Percent)		
Wheat	253,454	352,982	39.3	5.46	4.83	-11.5
Maize	210,074	317,303	51.0	3.29	4.41	34.0
Rice	119,971	155,031	29.2	3.97	3.80	-4.3
Barley	95,283	150,997	58.5	4.81	7.50	55.9
Millets	19,758	21,370	8.2	7.91	5.66	-3.2
Sorghum	40,233	53,386	32.7	5.22	5.70	9.2
Oats	49,035	47,600	-2.9	11.30	5.35	-52.6
Other Cereals	41,404	35,231	-14.9	4.57	9.31	103.7
Total Cereals	829,215	1,133,902	36.7	2.78	3.37	21.2

Source: Hazell (1984: Table 3).

[a] Does not include the People's Republic of China, because the early 1960s were a period of very unusual and extreme production instability in that country.

Table 3—Variability in export prices for wheat and rice in real terms, 1950-1979

	Wheat		Rice	
	Standard Deviation	Coefficient of Variation	Standard Deviation	Coefficient of Variation
1950-59	26.0	11.2	59.0	11.4
1960-69	7.0	3.6	89.0	17.5
1970-79	56.0	30.0	187.6	39.0

Source: Valdes (1984: Table 1).

Table 4—Comparison of growth of demand for agricultural commodities at different stages of development, hypothetical cases

Level of Development	Percent of Population in Agriculture	Rate of Population Growth	Rate of Per Capita Income Growth	Income Elasticity of Demand	Rate of Growth in Demand
Very low income	70%	2.5	0.5	1.0	3.0
Low income	60%	3.0	1.0	0.9	3.9
Medium income	50%	2.5	4.0	0.7	5.1
High income	30%	2.0	4.0	0.5	4.0
Very high income	10%	1.0	3.0	0.1	1.3

Source: Adapted from Mellor (1966).

Table 5—Net import and growth rates for imports and exports of major food crops[a] in developing countries,[b] 1966-70 and 1976-80 averages

Country Group	Net Imports 1966-70	Net Imports 1976-80	Annual Growth Rate 1966-70 to 1976-80 Exports	Annual Growth Rate 1966-70 to 1976-80 Imports
	(in million metric tons)		(percent)	
Developing countries	12.2	37.9	2.7	6.3
By region				
Asia	14.2	16.3	5.2	3.1
N. Africa/Middle East	4.8	17.1	1.3	11.1
Sub-Saharan Africa	(-1.3)	4.4	(-7.1)	9.2
Latin America	(-5.5)	0.2	2.7	8.4
By GNP per capita growth rate[c]				
Less than 1.0%	2.8	6.6	(-3.4)	4.4
1.0% to 2.9%	(-0.4)	(-7.3)	3.8	(-0.7)
3.0% to 4.9%	4.9	1.3	3.2	9.0
5.0% and over	5.0	17.0	(-0.9)	10.1

Source: Paulino (forthcoming).

Notes:
[a] Trade data include bran and cakes for feed use.
[b] Includes a total of 105 Asian, African, Middle Eastern, and Latin American countries. The People's Republic of China is included.
[c] Calculated on 1961-80 constant GNP figures.

Table 6--Average annual growth rates of production, area harvested, and output per hectare for major food crops in developing countries, 1961-1980

Country Group	Average Annual Growth Rates of Major Food Crops, 1961-1980 Production	Area Harvested	Output per Hectare
	(percent)	(percent)	(percent)
Developing countries	2.6	0.7	1.9
Asia (incl. China)	2.8	0.4	2.4
N. Africa & Middle East	2.5	1.1	1.4
Sub-Saharan Africa	1.6	1.5	0.1
Latin America	2.8	1.5	1.3

[a] Includes cereals, roots and tubers, pulses, and groundnuts. Rice is in terms of milled form.

[b] Annual growth rates of production may differ slightly from those shown in other tables, because of the differences in time period and because the data here exclude the outputs of bananas and plantains, for which estimates on the area harvested are not available.

4

Government Intervention, Financial Constraint, and the World Food Situation

Mathew Shane

"Man is not part of the landscape. He is the creator of the landscape." Bronowski, Ascent of Man

A fundamental question for economists is why some countries have highly successful economic development patterns while others do not. How can we explain why some of the outstanding success stories, such as Japan and Korea, have occurred in environments that were relatively devoid of national resources while other countries, with seeming abundance, have not succeeded? It is my central tenet that man--the creator of the landscape--explains the difference. He creates an economic environment that either encourages the efficient utilization of a country's resources and generates effective investment and trade patterns and thus high rates of economic growth or, alternatively, distorts the price and incentive system away from a country's comparative advantage and generates inefficient investments, restricted trade, and low rates of growth. In some countries, the institutional structure evolves to encourage enterprise and fosters the benefits of trade and specialization while, in others, it tends to separate the economy from the international market place.[1/2]

There is a reflection of this dichotomy in the world food situation. One set of countries implicitly tax their agricultural sector, discourage production and exports and has inadequate foreign exchange earnings to meet the required food gap with commercial imports. These countries are dependent on food aid to feed their people.

We refer to them as "food crisis" countries. Another set of countries has generated dynamic growth based on outward trade orientation. In these countries growth has led to a more rapid increase in food consumption demand than in productive capacity resulting in large growth in agricultural imports. They pay for these food imports by expanding nonfood exports. We refer to these countries as "growth importers."[3] This paper will define, identify, and compare the patterns of trade and development of the "food crisis" countries with that of the "growth importers." We are particularly interested in whether those countries exhibit distinctly different patterns. The identification of differences will provide a preliminary basis for developing an "optimum" strategy for food self-sufficiency.

The next section will briefly survey the world food situation in which, on average, there are growing food supplies but also significant numbers of countries in which food availability is declining. I will then identify the growth importers and food crisis countries and compare their patterns of trade, growth, and debt accumulation. The paper will conclude by suggesting ways for countries to achieve food self-sufficiency in the current world environment.

1. The World Food Situation

World food production has been rising in the postwar period (Figure 1) (Figures at end of chapter). It is generally assumed that the distribution of these production gains is widespread. Most of the gain in per capita food production, however, has been occurring in the developed countries (Figure 2) (Figures at end of chapter). Because consumption is growing slowly there, almost 70 percent of the growth in net agricultural exports over the period 1961-63 to 1981-83 has originated in those economies (Table 1) (Tables at the end of chapter). At the same time approximately 50 developing countries, with three-quarters of a billion population, are increasingly dependent on imports to adequately feed their growing populations. Some of these countries have, by deliberate policy choices, decided to trade for food imports, while others, largely concentrated in Africa, have shown declining patterns of per-capita food production (Figure 3) (Figures at end of chapter).

Declining per-capita food production is not, by itself, a serious problem for developing countries. With world food production increasing, one development strategy could be to reduce resources to agriculture, where the returns might be low, and export other products with a higher return to generate the foreign exchange for food imports. This is, in fact, what one set of developing countries has done. Another set has followed policies to discourage food and agricultural production but has not developed the alternative export industries; instead, they paid for their excess of imports over exports by borrowing. These countries now find themselves faced with substantial long-term international debt that constrains their ability to pay for food imports.

2. Agricultural Growth Market Countries

Most of the increase in agricultural trade is accounted for by a relatively few countries. Only twenty-one countries account for more than 87 percent of the growth in net agricultural imports over the 20 year period, 1961-63 to 1981-83 (Table 2) (Tables at the end of chapter). Fifteen countries account for almost 94 percent of the growth in net exports (Table 3) (Tables at the end of chapter).

We define an agricultural growth importer as any developing market economy that accounted for at least 1 percent of the growth in world net agricultural imports between 1961-63 and 1981-83 and whose imports are principally commercial in nature. The growth market set of countries is given in Table 4 (Tables at the end of chapter). Four of them are oil Exporters, i.e., Saudi Arabia, Mexico, Nigeria and Venezuela. Two of these countries, Mexico and Nigeria, currently face severe debt repayment problems.

3. Food Crisis Countries

We define a food crisis country by the fact that its per capita food production is declining, its average per capita consumption is below accepted nutritional standards, and it lacks sufficient foreign exchange or international borrowing capacity to commercially import required food deficits.[4] A country in this situation would have difficulty responding to production shortfalls. The situation is even more critical for those countries whose per capita food consumption is below recommended levels, as well as declining.

Applying these criteria to a set of 106 developing countries, we identify 12 countries that currently appear to be faced with the most serious potential food crisis (Table 5) (Tables at the end of chapter). These countries were, on average, net agricultural Exporters over the period, although declining ones. The countries summarized in Table 5 will be our comparison group to those in Table 4.

We turn now to an analysis of the contrasting development patterns of the two sets of countries.

4. Measure of Government Policy Intervention-Nominal Rate of Protection

One measure of government intervention in the agricultural sector is the degree to which domestic price facing producers deviates from its border price for a specific imported item.[5] This measure is frequently referred to as the nominal rate of protection. If the hypothesis about policy differences has validity, then we should observe substantial and consistent differences between the growth importers producer-import price ratio and that for the food crisis countries. If the governments of food crisis countries have implicitly taxed their agricultural sector by consistently offering relatively low food prices to their producers, and if the growth importers have followed a more outward oriented policy towards their agricultural sector, one would expect the price ratio to be significantly lower for the food crisis group than for the growth importers.[6/7] Since the interventions occur with regard to individual crops, and since the grains and Oilseeds are the major trade crops, we will concentrate our investigation on them. Using trade and production weighted prices, we derive price indices for 1967-82.[8] As expected, the ratios for the growth importers are uniformly higher than for the food crisis countries. This is true for the grains and Oilseeds combined (Figure 4) (Figures at end of chapter), as well as for the food grains (Figure 5) (Figures at end of chapter), coarse grains (Figure 6) (Figures at end of chapter) and the Oilseeds (Figure 7) (Figures at end of chapter) separately.

For the grain prices (Figure 5-6), the ratio declines over time for the food crisis countries and increases for the growth importers. There also seem to be two distinct patterns over time for both sets of countries, an upward pattern from 1967-71 which seems to continue in 1980-82, and a dramatic cyclical pattern from 1971-80. Undoubtedly this is an indication of the adjustment in the grain markets from the oil shock and the Russian wheat deal. There is an increasing time pattern in the oilseed price ratio (Figure 7) for both the food crisis and growth importers. Furthermore, there does not appear to be the substantial cyclical patterns observed in the grain markets.

5. Patterns of Real Exchange Rates

A second measure of government intervention policy in the foreign trade sector is the degree to which countries have allowed their currencies to become over or undervalued. This can be measured by the real exchange rate. We have developed group indices by taking individual country indices and aggregating these with agricultural trade weights. We set 1980 as the base year.

Two distinct patterns emerge over the period 1961-84 (Figure 8) (Figures at end of chapter). From 1961-80, there is a definite tendency for the real exchange rate indices to decline for both the food crisis and selected importer countries, while in the period 1980-84, there is a tendency for them to rise. In both the former and latter period, the movement of the food crisis countries exceeds that of the selected importers.

A decline in the real exchange rate index (measured as local currency over US$) implies a real appreciation of the currency. This would tend to encourage imports and discourage exports, a pattern consistent with what indeed took place. The food crisis countries were tending to use the overvaluation of their exchange rate as a means for providing cheap imports for their urban sectors. This also, however, had the effect of discouraging exports and the export sector. All of which will lead to a situation in which large current account deficits will become pervasive. The almost fourfold rise in the real exchange rate index of the food crisis countries since 1980, a period of international financial constraint and adjustment, provides an indication of the enormous degree to which exchange rates were overvalued in these countries.

6. Outward or Inward Orientation

The period of the 1960's and 1970's were periods of great expansion in world trade. In all but three years, world trade expanded faster than world Gross Domestic Product (GDP). Indeed, trade grew approximately 1.6 times that of GDP, on average for the developing countries. Those countries that utilized trade as an engine of growth tended to be more successful than those that did not. In this regard, it is interesting to look at the patterns of exports to GDP for the two groups (Figure 9) (Figures at end of chapter). Over the period of 1961-84, the growth importers were increasingly outward oriented and display a pattern of increasing exports to GDP, while the food crisis countries were increasingly inward oriented and display a decreasing pattern. Thus, during this period when there was a great opportunity to utilize trade as an engine of growth and development, the growth importers took full advantage of this opportunity while the food crisis countries increasingly withdrew from world markets.

This approach to trade is also clearly mirrored in patterns of trade between the two groups (Figure 10-11) (Figures at end of chapter). Whereas the growth importers tended to run a consistent trade surplus over the entire period, the food crisis countries have a pattern of worsening deficits and lagging exports.

7. Per Capita Incomes

These pronounced differences in policies regimes have resulted in dramatically different patterns of per capita income growth from 1960-84 (Figure 12) (Figures at end of chapter). While the growth importers have shown consistent increases in real per capita incomes from an average of approximately $2700 to more than $6000, the per capita incomes of the food crisis countries have been stagnant at around $400.[9]

This is brought out clearly in the smoothed pattern of annual changes in per capita income for the two groups (Figure 13) (Figures at end of chapter). While both groups have shown a pattern of secular decline in the increments, the food crisis countries have actually had negative income growth over the last ten years of the period. Indeed, with the exception of 1982, the selected importers always had higher average growth than did the food crisis countries. The dramatic shift in exchange rates in 1983-84 might be having a positive effect on growth in the food crisis countries starting in 1984, but a new long-term pattern must still be demonstrated.

8. The International Debt Picture[10]

The statistics on debt are unfortunately not as complete as that for the other statistics we have utilized, so we will have to restrict ourselves to an analysis of the period 1973-84. The period 1973-80 was one of very substantial debt accumulation. It was also one of very substantial growth and prosperity in the developing countries. It is an irony of the best of times that there is always a set of countries, which, by poor economic planning and policy, generate for themselves the worst of times. This, indeed, seems true for the food crisis countries.

The accumulation of debt was fundamentally related to the international balance-of-payments disequilibrium generated after the first oil shock of 1973-74. The fourfold increase in oil prices by members of OPEC dramatically altered payment flows and the international financial environment, initiating the process by which significant debt was accumulated.

The industrial market (OECD) countries employed easy monetary policies both before and after the first oil shock. The change in trade flows and expansionary monetary policy in the OECD nations generated a liquidity previously unavailable to the international system. International bankers recycled this liquidity in the form of petrodollars, by beginning a massive lending program first to oil importing countries, but later to all developing countries. These bankers anticipated high returns on investments.

International borrowing, in and of itself, is neither good nor bad for subsequent development. If the extra resources utilized for investments led to a stream of foreign income more than sufficient to repay the loans, then accumulating debt has facilitated development. On the other hand, if borrowed capital is used to increase subsidized consumption and other policies of market distortions or invested in government projects and government companies with low or negative returns, then international borrowing becomes an impediment to future economic growth.

The second oil shock of 1979-80 set the stage for the world recession of 1980-83. The second oil increase turned out to be far more important than the first one because of the very different policy responses in the developed world. The approach to the 1973-74 increase had been to find ways to recirculate petrodollars and to accommodate the jump in energy costs. The response to the 1979-80 increase, however, was for the major industrial countries to simultaneously undertake contractionary monetary policies. The world inflation that was initiated by the 1973-74 oil increase, but by no means limited to it, proved unacceptable to the West, and Western countries felt that only traditional measures could deal with the inflation. The sudden lowering of monetary growth slowed the world economy, raised real interest rates, and made the debt a burden for those countries unwilling or unable to respond to the change in world environment.

The patterns of debt accumulation for the growth importers and the food crisis countries reflects the overall pattern for the developing countries, but with some special features (Figure 14) (Figures at end of chapter). The growth importer's medium and long term international debt has increased from $33 billion to over $140 billion while the food crisis countries debt has increased from almost $10 billion to around $23 billion in 1984.

This pattern of debt accumulation is reflected in the annual percentage change in debt (Figure 15) (Figures at end of chapter). Both groups show a pattern of declining debt accumulation over time. The growth importers have been accumulating debt at a faster rate than the food crisis countries. Moreover, the 14 percent rate of accumulation between 1979-84 is substantial compared to all developing countries and the food crisis countries. The continued accumulation of debt in the current economic environment implies a low risk evaluation by the financial community of this group of countries. These countries utilizing international borrowing for profitable ventures do not appear to be faced with a significant financial constraint.[11]

Both the food crisis countries and growth importers have had patterns of debt accumulation implying rising debt to GDP ratios, particularly since 1980-81 (Figure 16) (Figures at end of chapter). However, the food crisis countries have a higher ratio and one that increases faster. The really significant pattern indicating the difference in the constraint is the ratio of debt to exports (Figure 17) (Figures at end of chapter). The selected importers have a relatively low, flat and stable pattern, whereas the food crisis countries have a high rapidly rising one, particularly since 1980.

Thus, although the growth importers have been accumulating debt in both larger amounts and higher rates, it is the food crisis countries that are debt constrained. This tends to substantiate the assertion that it is the use of the credit rather than the accumulation per se that is important. Countries which support government intervention policies through international borrowing will rapidly become debt constrained.

9. Patterns of Agricultural Trade

Since it is the growing reliance on food imports which is the common characteristic of both the growth importers and the food crisis countries, it is appropriate to end the analysis with an evaluation of how this import reliance has come about.

The growth importers have both rising agricultural imports and exports with imports rising more rapidly than exports. Together this implies that they are substantial and rising net importers (Figure 18) (Figures at end of chapter). The food crisis countries historically have been net agricultural Exporters, although net food importers (Figure 19) (Figures at end of chapter).[12] However, while imports have been rising, exports have been falling so that by 1984, they were for the first time over the historical period 1961-84 net agricultural importers. This is consistent with a policy mix which taxes agricultural producers and subsidizes consumers as indicated by their nominal rate of protection.

10. Conclusions

One group of countries which have maintained food policies which favor consumers over producers--so called cheap food policies--generate an outcome, not surprisingly in which food becomes a scarce good. A cheap food policy has, in essence, become expensive.

Another set of countries has created an environment where the rewards to producers are high and international competitiveness is maintained. Since these countries have decided that their comparative advantage is outside of agriculture, they are, as a group, significant agricultural growth market countries. However, because they have generated the necessary export earnings, the importation of the required agricultural product has not presented any problem for them.

Thus, we return to the question of the best strategy a country can take to maintain food security. From the law of comparative advantage we know that a country should tend to specialize in those commodities or products in which it has a relative advantage. By so doing, a country can take advantage of the gains from trade. Attempting to maintain food self-sufficiency through an inward orientation, implies that countries are giving up their trade benefits and encouraging their farmers to produce the full spectrum of agricultural commodities.[13] Encouraging the production of relatively low productivity commodities led to a lowering of overall agricultural output. Trying to produce more results in less. Inward orientation is, therefore, almost certain to fail to provide the food self-sufficiency. Instead, it is more likely to generate food shortages that can only be met through grants from donor countries.

Footnotes

[1] Unfortunately Adam Smith's "invisible hand" does not occur by itself. We must create the appropriate environment. In some societies the rules of conduct are such that the individual objectives are consistent with the social objectives. In others, however, individual objectives conflict with social objectives in varying degrees. Therefore, individual maximization will not lead to social optimization. This is particularly true if there is a large number of externalities or if a substantial amount of resources are devoted to rent seeking.

[2] Most countries fall in between these two extremes. Indeed, of the total number of developing countries only 10 percent can be characterized as highly successful 10 percent are food crisis countries.

[3] In the following section, we will define more precisely what we mean by the "food crisis" and "growth importer" countries.

4 For an analysis of food crisis countries: Yetley, Merv. and Mathew Shane. Food Deficit Countries Under Financial Constraint. FAER forthcoming, U.S. Department of Agriculture, Economic Research Service.

[5] There are a number of measures of government intervention in foreign trade market, of which this is one. Another measure of the degree of separation of the domestic and foreign market is the price transmission elasticity. For a discussion of this see: Roe, Terry, Mathew Shane and De Huu Vo. Price Responsiveness of International Grain Markets: The Impact of Government Intervention on Import Elasticities, Technical Bulletin 1720, U.S. Department of Agriculture, Economic Research Service, June 1986.

[6] Although it is beyond the scope of this paper, there is considerable evidence that many low income developing countries intervene in their food sectors in a variety of ways to lower the price of food to their consumers. See the ERS/U. of Minn.. Food Policies in Developing Countries, Foreign Agricultural Report No. 172, Econ. Res. Serv., U. S. Dept. of Agr., September 1983. Countries use a variety of means to artificially depress food prices. In some instances it is done by setting up a

state monopoly to control the trade and marketing of key commodities. In other instances it is accomplished by setting up administered prices. The rationale for depressing producer prices is to provide low cost food to urban consumers. This is based on the fact that the food component of the budget in low income countries can be as high as 50 percent on average and as high as 90 percent for the lowest income group of the population. Furthermore, since food makes up such a significant component of the budget, the newly emerging industrialists support large consumer subsidies since it also tends to lower wage costs.

[7] The term outward oriented was coined by Bela Balassa. It refers to countries that maintain unbiased or free trade orientation.

[8] The price data is derived from FAO Production and Trade Yearbook tapes by dividing the annual dollar value of production imports by the wheat equivalent quantity of production or imports and aggregating by quantity weights.

[9] One could argue that this is an unfair comparison since the growth importers are middle and upper middle income countries while the food crisis countries are lower income countries. However, the per capita income of Korea in 1960 was not distinctly different from those of the food crisis countries. Indeed, one would not have selected Korea to be one of the growth miracles of the 1960's and 1970's in 1960. It was the dramatic change in national policy from "inward" to "outward" orientation including the freeing up of the capital and foreign exchange market which led to the spectacular growth observed over the period.

[10] For a more complete discussion of this issue see: Shane, Mathew and David Stallings. Financial Constraints to Trade and Growth: The World Debt Crisis and its Aftermath, FAER 211, U.S. Dept. of Agr., Econ. Res. Serv., December 1984 and Shane, Mathew and David Stallings. Financial Constraints to Trade and Growth: The Debt Crisis and Its Resolution, U.S. Department of Agriculture, Economic Research Service, forthcoming.

[11] The size of debt is not a measure of financial constraint. Korea has one of the largest international country debt at $40 billion but because of its outward orientation, excellent investment practices and rapid adjustment to changing international conditions continue to grow rapidly throughout the 1980's. For an analysis of their early adjustment response see: Aghevli, Bijan and Jorge Marquez-Ruarte. A Case of Successful Adjustment:

Korea's Experiences During 1980-84, Occasional Paper 39, International Monetary Fund, Washington, D.C., August 1985.

[12] The food crisis countries have been exporting largely non-food agricultural exports such as coffee, tea, rubber, palm oil, fibers and tropical fruits.

[13] This typically is done by providing relatively high price supports for commodities in which the country has no comparative advantage and relatively low price supports for those commodities in which the country has a comparative advantage. An example of this is the case of Tanzania. See Gerrard, Christopher and Terry Roe. "Government Intervention in Food Grain Markets: The Case of Tanzania," Journal of Development Economics, Vol. 12, No. 1, 1983, pp. 109-32.

Table 1—Distribution of Growth of Net Agricultural Imports and Exports Among Industrial Market, Planned, and Developing Countries, 1961-63 to 1981-83 (in percent)

Country/Group	Total Agricultural Trade	Grains and Oilseeds d/	Food Grains a/	Coarse Grains b/	Oilseeds c/
	Share of Growth in Net Imports				
Industrial Market	24.60	19.15	3.34	23.00	43.60
Planned Economies	23.49	33.63	33.68	28.19	16.83
Developing	51.91	47.21	62.99	48.81	39.56
Middle Income	23.62	25.91	19.24	35.22	18.82
Low Income	20.36	18.46	41.17	9.60	20.45
Oil Exporters	7.93	2.84	2.58	3.99	0.29
	Share of Growth in Net Exports				
Industrial Market	69.49	75.76	-74.35	81.34	66.68
Planned Economies	4.07	0.81	5.42	1.17	0.11
Developing	28.08	20.25	20.23	17.49	33.18
Middle Income	17.47	16.46	10.28	13.80	31.97
Low Income	10.61	3.78	9.95	3.70	1.21
Oil Exporters	0.00	0.00	0.00	0.00	0.00

a/ Includes wheat, wheat flour and rice.

b/ Includes maize, barley, sorghum and millet.

c/ Includes soybeans, soybean oil, soybean meal, groundnuts, groundnut oil, groundnut meal.

d/ The sum of a/, b/, and c/.

Source: U.S. Dept. of Agriculture, Economic Research Service data base.

Derivation: Countries are grouped according to whether they are net importers or Exporters and as to whether they are industrial market, planned or developing countries. The change in net imports (exports) of the group is then calculated as a percent of the overall change in net imports (exports).

Table 2--Real Value of Agricultural Net Imports and Growth in Agricultural Net Imports, Major Importers, 1961-63 to 1981-83

Country/ Market Category	------- Net Imports ------		:	--Percentage of Net Imports--		
	1961-63	1981-83	Change 1961-83	1961-63	1981-83	Change 1961-83
	Millions of Constant 1974-76 US$			----------Percent----------		
USSR	511	12,637	12,126	1.4	14.6	19.4
Japan	4,396	14,011	9,614	12.3	16.2	15.4
Italy	2,139	5,877	3,739	6.0	6.8	6.0
Saudi Arabia	179	3,370	3,191	0.5	3.9	5.1
Iran	53	2,795	2,741	0.1	3.3	4.4
Korea Republic	224	2,727	2,503	0.6	3.1	4.0
Egypt	-408	2,072	2,480	-1.3	2.4	4.0
Mexico	-1,015	1,279	2,294	-3.3	1.5	3.7
Nigeria	-638	1,466	2,103	-2.1	1.7	3.4
Algeria	100	2,152	2,052	0.3	2.5	3.3
Iraq	176	2,026	1,850	0.5	2.3	3.0
Hong Kong	702	1,944	1,242	2.0	2.2	2.0
China (PR)	714	1,939	1,225	2.0	2.2	2.0
Taiwan	-229	988	1,217	-0.7	1.1	1.9
Venezuela	168	1,256	1,089	0.5	1.5	1.7
Poland	252	1,254	1,002	0.7	1.4	1.6
Germany (F.R.)	7,403	8,388	985	20.7	9.7	1.6
Libya	68	1,003	935	0.2	1.2	1.5
Portugal	237	1,114	877	0.7	1.3	1.4
Spain	295	927	632	0.8	1.1	1.0
Morocco	68	694	626	0.2	0.8	1.0

Source: Cesal, L. et al. Agricultural Growth Market in a World Economy. FAER forthcoming, U.S. Department of Agriculture, Economic Research Service.

Table 3--Real Value of Agricultural Net Exports and Growth in Agricultural Net Exports, Major Exporters, 1961-63 to 1981-83

Country/ Market Category	------ Net Imports ------			--Percentage of Net Imports--		
	1961-63	1981-83	Change 1961-83	1961-63	1981-83	Change 1961-83
	Millions of Constant 1974-76 US$			---------Percent---------		
United States	3,699	21,506	17,807	12.0	29.4	32.9
Brazil	2,713	6,621	3,907	8.8	9.0	7.2
Netherlands	378	4,706	4,328	1.2	6.4	8.0
Thailand	821	3,083	2,262	2.7	4.2	4.2
France	-1,622	2,974	4,596	-4.5	4.1	8.5
Argentina	2,397	4,567	2,169	7.8	6.2	4.0
Canada	1,409	4,267	2,859	4.6	5.8	5.3
Australia	3,755	5,652	1,898	12.2	7.7	3.5
Hungary	-62	1,217	1,278	-0.2	1.7	2.4
Malaysia	255	1,551	1,295	0.8	2.1	2.4
Denmark	1,249	2,075	826	4.0	2.8	1.5
India	-191	609	800	-0.5	0.8	1.5
Turkey	406	1,591	1,184	1.3	2.2	2.2
New Zealand	1,280	2,083	802	4.1	2.8	1.5
Yugoslavia	-52	-25	27	-0.1	.0	.0

Source: Cesal, L. et al. Agricultural Growth Markets in a World Economy. FAER forthcoming, U.S. Department of Agriculture, Economic Research Service.

Table 4—Growth Market Countries Based on Increases in Commercial Agricultural Net Imports, 1961-63 - 1981-83

Japan	Hong Kong
Italy	Taiwan
*Saudi Arabia	*Venezuela
Korea Republic	Portugal
*Mexico	Spain
*Nigeria	
Algeria	

* These are oil Exporters.
Source: Applying the growth market criteria to the countries identified in Table 2.

Table 5—List of Potential Food Crisis Countries

Bangladesh	Peru
Chad	Sierra Leone
Ethiopia	Tanzania
Kenya	Togo
Mozambique	Uganda
Nepal	Zaire

Source: Taken from Yetley, Merv and Mathew Shane, Food Deficit Countries Under Financial Constraint. Forthcoming, U.S. Department of Agriculture, Economic Research Service.

Figure 1

Food Production Indices, World

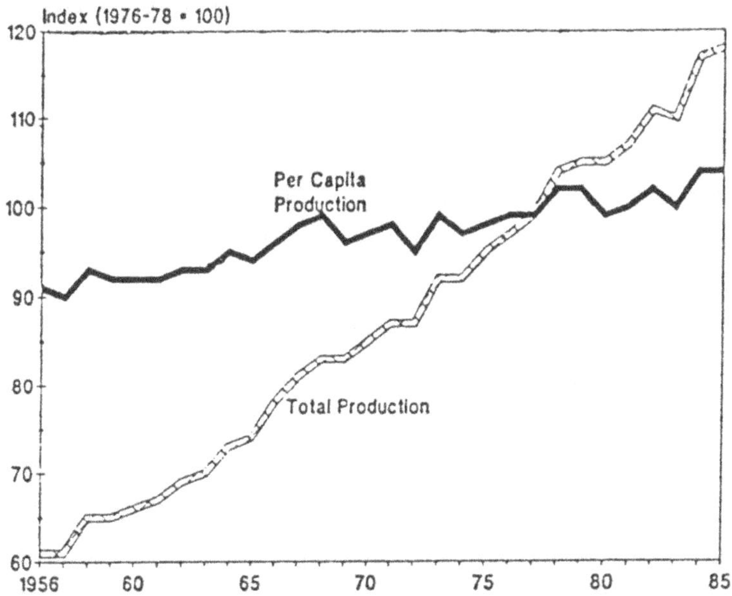

Figure 2

Food Production Indices, Developed Countries

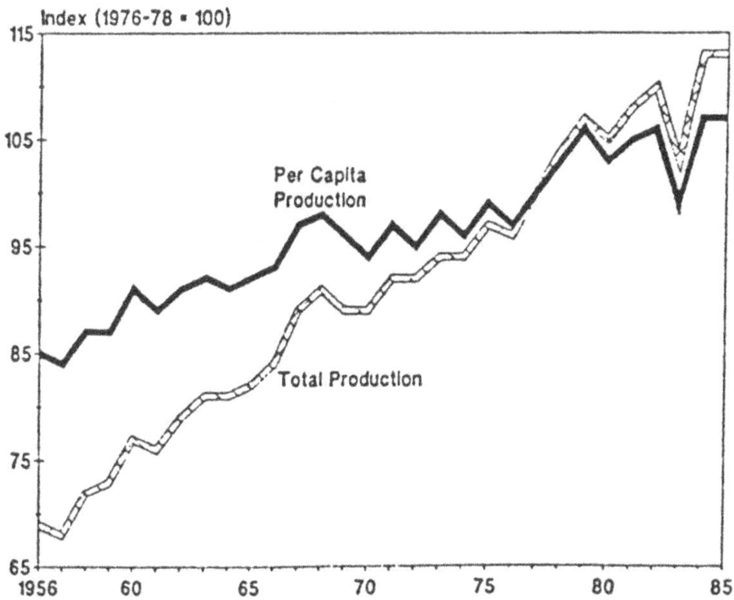

Figure 3

Food Production Indices, Africa less Republic of South Africa

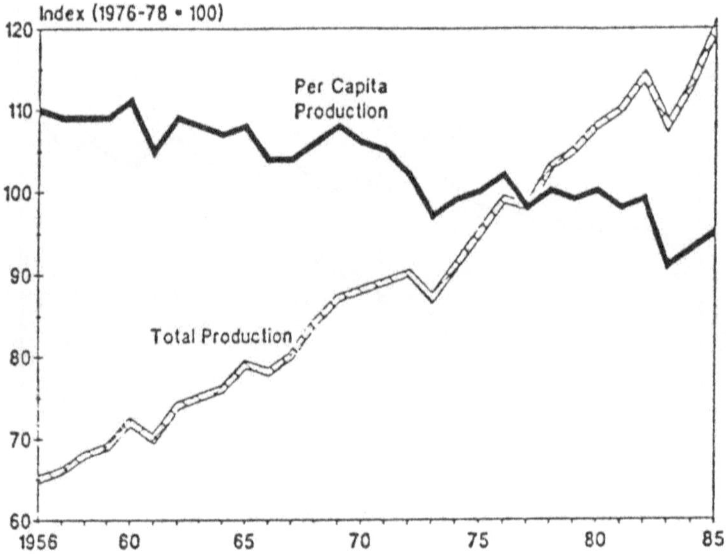

Figure 4

Grains and Oilseeds Producer to Import Price Ratio

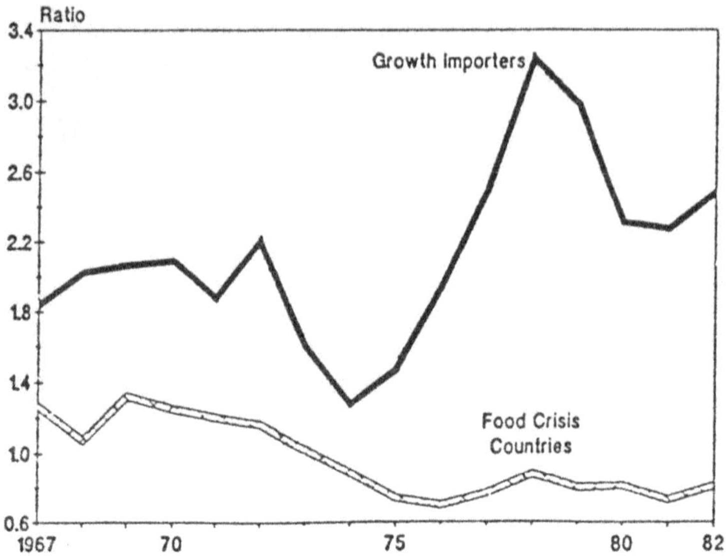

Figure 5

Food Grains
Producer to Import Price Ratio

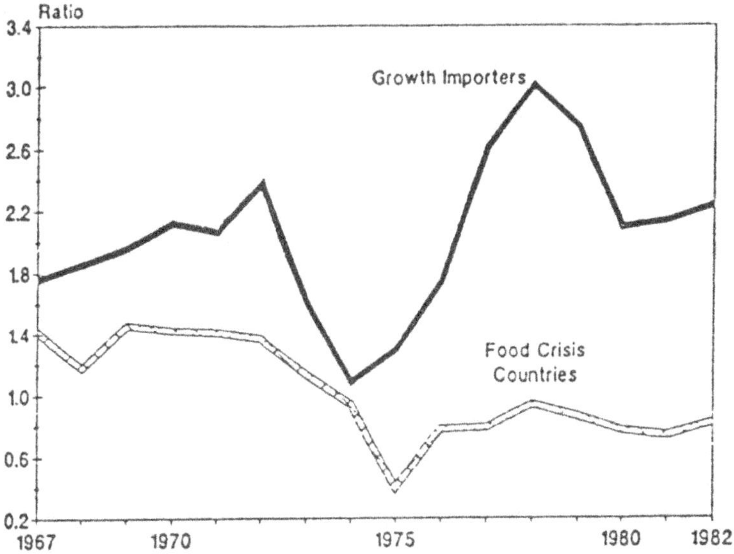

Figure 6

Coarse Grains
Producer to Import Price Ratio

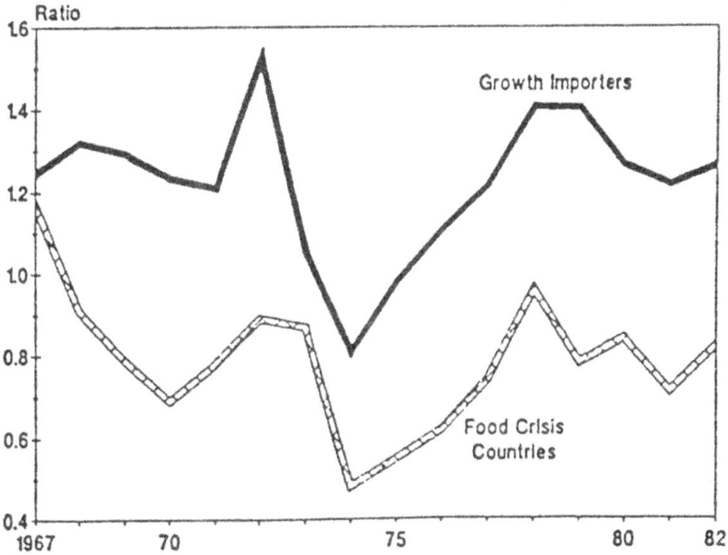

Figure 7
Oilseeds
Producer to Import Price Ratio

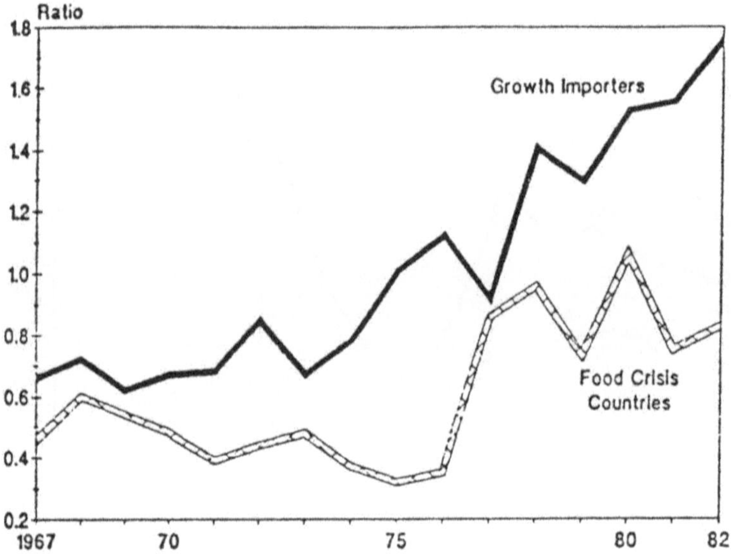

Figure 8
Exchange Rate Index
Trade Weighted

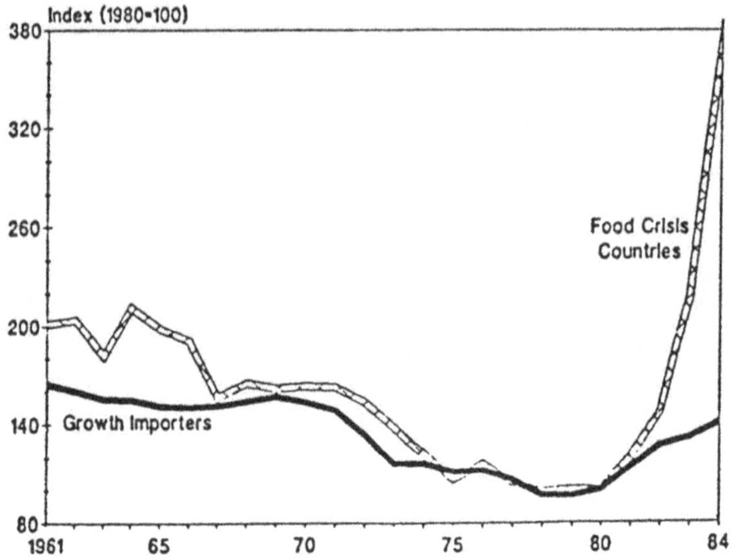

Figure 9
Export to GDP Ratio

Figure 10
Growth Importers
Trade in Goods and Nonfactor Services

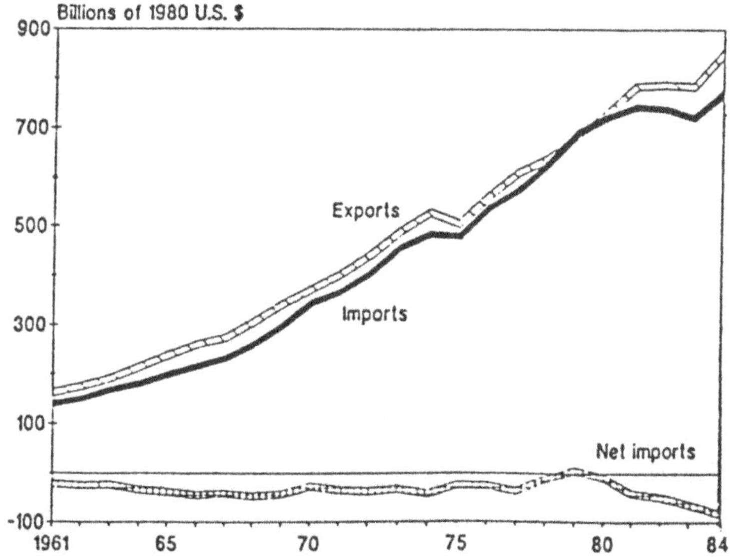

Figure 11
Food Crisis Countries
Trade in Goods and Nonfactor Services

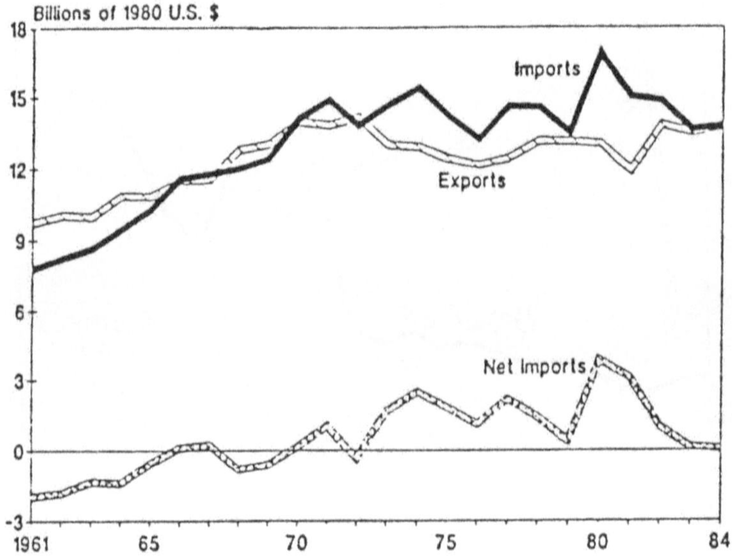

Figure 12
Per Capita GNP

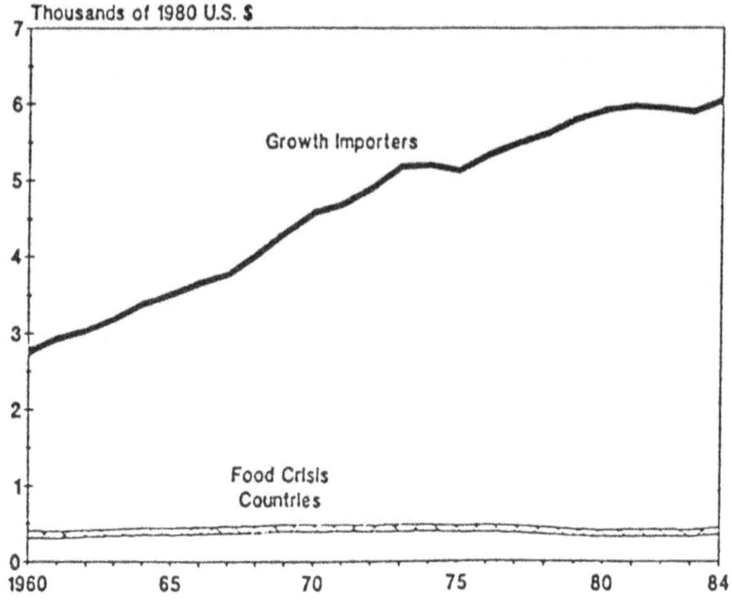

Figure 13

Change in Real GDP
Per Capita 5 Year Moving Averages

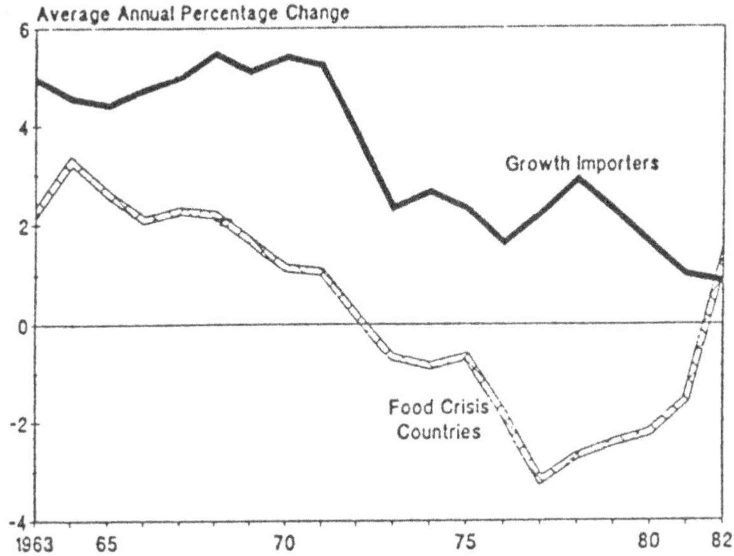

Figure 14

Real International Debt
Medium and Long Term

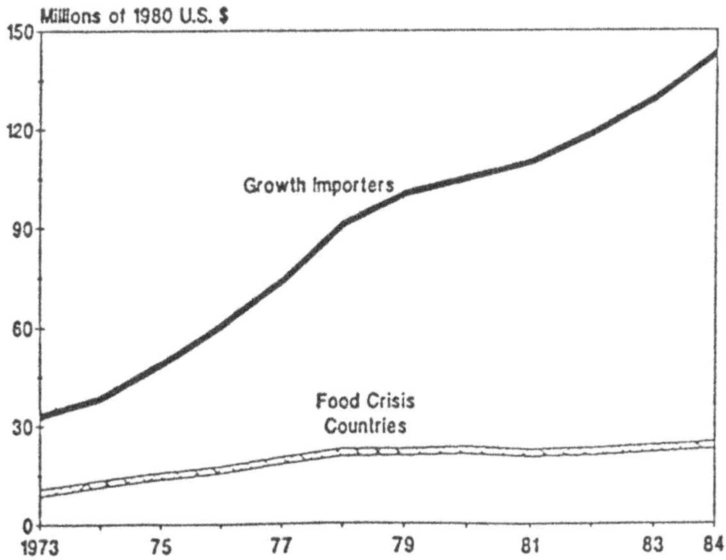

Figure 15
Annual Change in Debt

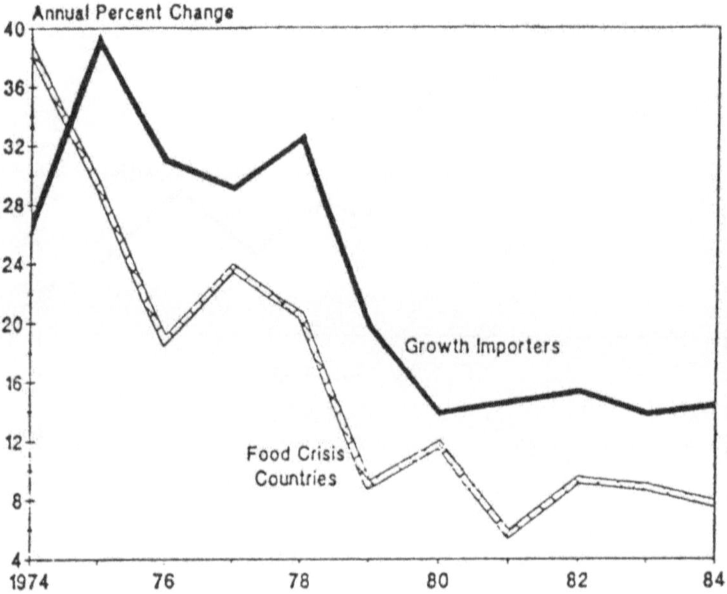

Figure 16
Debt to GDP Ratio

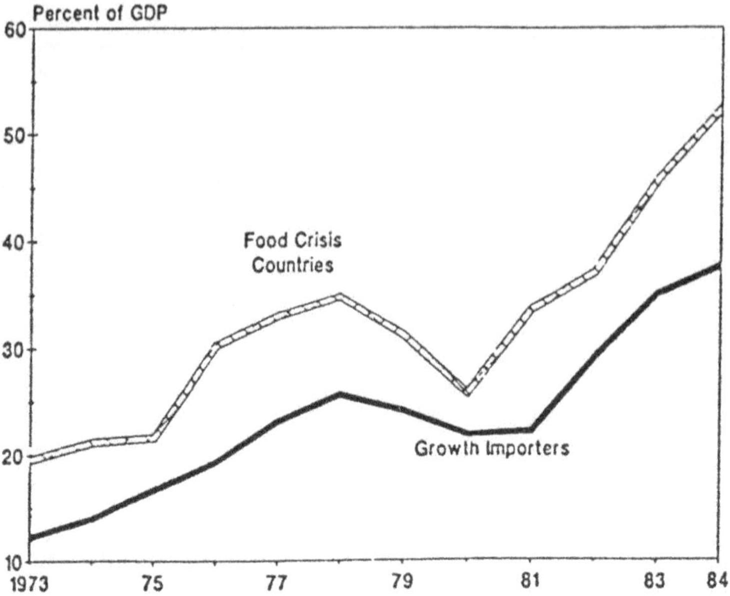

Figure 17

Debt to Export Ratio

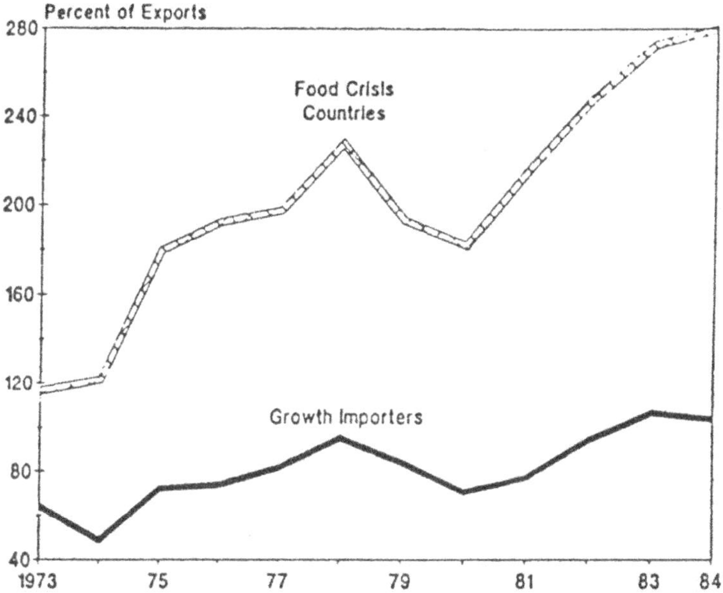

Figure 18

Growth Importers Real Value Total Agricultural Trade

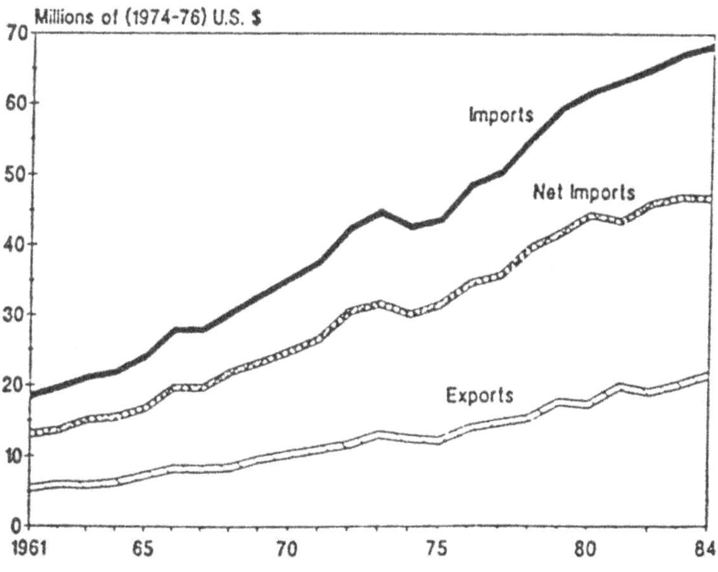

Figure 19: Total Agricultural Trade

Food Crisis Countries, 1961–84

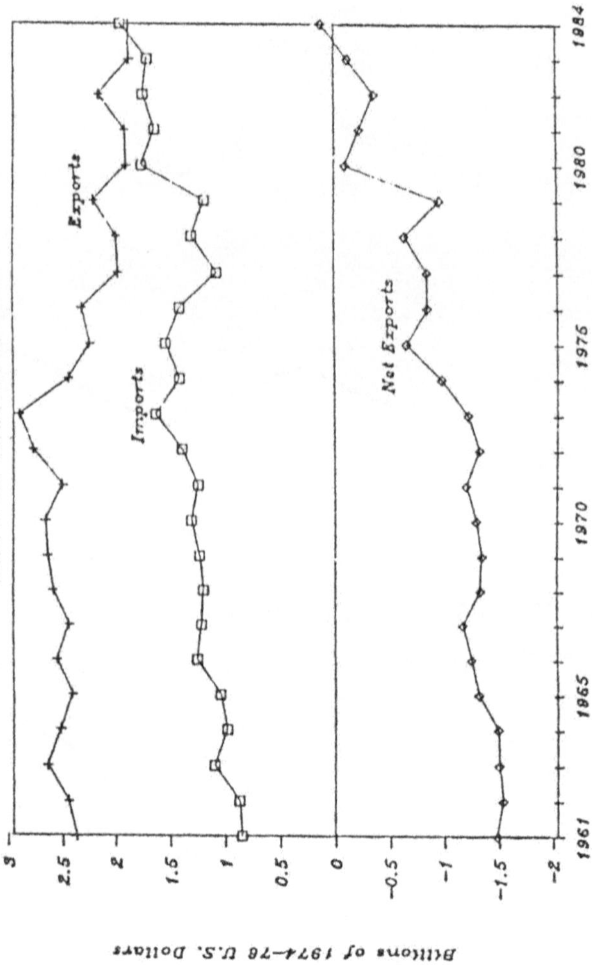

5

Agricultural Policies in Developing Countries: The Transfer of Resources from Agriculture

Terry L. Roe

I. INTRODUCTION

This paper focuses on interventions by governments of developing countries that tend to distort economic incentives and, in particular, to transfer resources from agriculture. Generally, as agriculture is modernized and an economy develops, labor moves out of agriculture at the same time that capital deepening[1] occurs in the form of investment in land and human capital and the adoption of chemical, biological, and mechanical technology. Since markets work imperfectly in the provision of new agricultural technology, rural infrastructure, education and other services that markets depend on, government programs and projects to remedy these imperfections are often socially profitable. Many countries, however, have pursued policies that have subsidized food to urban consumers while protecting inefficient domestic industry from foreign competition. The result has been a regime of domestic prices that depart from those that would otherwise reflect the true opportunity cost of a nation's resources, an underinvestment in areas where markets work imperfectly, and a movement of resources out of agriculture and into those sectors of the economy that appear--artificially, because of these distorted prices--more profitable. Hence, whether governments should intervene is not at issue; instead, at issue is the form of intervention.[2]

Evidence of interventions that have distorted agricultural prices (i.e., a regime of prices that do not reflect the true opportunity cost of resources employed in the sector), and implicitly taxed the sector, has been reported in numerous studies. For instance, Bale and Lutz found agriculture in four developing countries to be heavily taxed, with direct welfare losses from distortions in agricultural prices ranging from 10.6 percent of GNP in Egypt to 1.5 percent of GNP in Argentina. Similar price distortions were observed in a study of Tanzania (Gerrard and Roe). Bautista found that export taxes and an overvalued exchange rate were the major factors contributing to depressed rice exports and low producer prices for rice in the Philippines. In Peru, the subsidizing of consumer prices led to a decline in producer prices that were further depressed by the country's overvalued exchange rate (Orden, et al.).[3]

Interventions in agriculture, however, are not alone in distorting that sector's economic incentives. In many countries, interventions in agriculture are accompanied by interventions in the urban-industrial sector that make this sector appear relatively more profitable. The interventions thereby, serve both to push and pull resources away from agriculture. Moreover, there appears to be a pattern to intervention among developed and developing countries that suggests that socially inefficient interventions might be explained by a more fundamental underlying process than simple policy mistakes.

The paper is organized as follows. The next section is composed of two parts, the first of which takes a closer look at government intervention in foreign trade markets, with emphasis on agricultural commodities. The second part focuses on interventions in agricultural procurement and marketing activities. The next two sections consider some of the macroeconomic effects of these interventions and how they can increase a country's susceptibility to shocks in world markets. Then the question is considered: If social economic losses from these interventions are large, why are they pursued? The paper is concluded with comments on the problems of removing distortions.

II. THE PROCESS OF TRANSFERRING RESOURCES FROM AGRICULTURE

Agricultural policy in most countries is interwoven with policies in other sectors of the economy. Discussion is therefore facilitated by considering how the intervention seems to take place in a "typical" country. We illustrate by referring to the experience of selected countries.

Since agriculture is the dominant sector in most developing countries, it is often an important resource base that policy makers might tax to finance public investment both in agriculture and other sectors. It is difficult to tax this resource because of the large number of spatially dispersed and heterogeneous component households and the lack of infrastructural development that characterizes agriculture in most developing countries. Consequently, resources are often extracted from agriculture through the use of indirect methods, such as interventions that lower the price of agricultural products. These interventions give rise to what we refer to as implicit taxes.

Policy makers are tempted to use indirect methods because food is often a large part of the household budget. The share of total income spent on food by low income urban households in developing countries may exceed 60 percent of total household expenditures. Hence, interventions that lower the price of food are effective in transferring resources to the urban sector because the lower prices may create significant increases in the real wages of urban workers. While these kinds of interventions tend to distort economic incentives, they are popular with urban households and industrialists for reasons we shall discuss later.

Agriculture can clearly benefit from a transfer of resources that are reinvested by the public sector. It is well known that markets alone are not effective in producing yield-increasing technologies without public sector support. Rural households have benefited from public investments in yield-increasing biological and chemical technology, irrigation facilities, and flood control. And, they have benefited from nonagricultural investments that serve to improve communications, lower transportation costs, and provide public utility services, improved educational opportunities, and other human capital-augmenting investments.

The social profitability of public investment in agriculture relative to that in other sectors of the economy depends, in part, on the level of per capita income. As per capita income increases beyond the level required to meet basic food needs, the incremental demand for nonagricultural goods and services tends to exceed the incremental demand for food. Hence, all else the same, the returns to resources employed in the nonagricultural sector of the economy tends to exceed those employed in agriculture. Likewise, the social return to incremental public investments in the nonagricultural sector tend to exceed those in agriculture. In many developing countries, however, a large fraction of the population is still struggling to meet basic food needs so that an increase in income gives rise to a greater increase in the demand for food than for nonagricultural goods and services. Then, it is possible for the social return to an incremental public investment in agriculture to exceed the return in other sectors of the economy.

A recent study of 4,000 households in the Dominican Republic reveals the magnitude of demand elasticities for food compared with that of nonfood products and underscores the dependence of these elasticities on income levels. As shown in Table 1 (Tables at the end of chapter) total food-expenditure elasticities[4] tend to decline with increases in household income. For both rural and urban households, however, these elasticities remain fairly large through medium income levels. The share of total household expenditures on food is over 50 percent for all households, except those in the urban high-income group. For an economy dominated by rural low and medium-income households, expenditure elasticities for food actually exceed those for housing, clothing, and other nonfood goods and services. The elasticity of the industrially produced commodities, clothing, and other nonfood goods and services, tends to exceed food expenditure elasticities only as households move into the highest income group. In the case of urban high-income households, expenditures on nonfood items exceed 60 percent of total expenditures. Hence, the return to private and public investments in agriculture may exceed returns in the nonfood sector when, in the early stages of development, the rising demand for food exceeds the growth in demand for nonfood goods and services.

The two fundamental issues that arise are the appropriate rate of resource transfer within and out of agriculture and the means to carry out the transfers. The socially optimal rate of transfer, for the most part, depends on the discounted expected social returns to capital in market and nonmarket activities, which, in turn, depend on the expected growth in final demand and in foreign trade opportunities. Approximating the socially optimal rate of transfer from the agricultural sector to just the nonmarket sector would seem to be a heavy burden for even the most sophisticated planning process in developing countries.

The means of resource transfer is the critical issue on which I want to focus. Relying on markets to transfer capital from agriculture means giving rural households adequate incentives to accumulate and hold savings in forms that can be used to fund nonagricultural-type investments. Modern rural capital markets often incur high transaction costs because of the large number of small transactions. While some success at establishing such capital markets has been realized (Hayami and Ruttan, pp. 399-403), their performance is conditioned by a country's monetary, fiscal, and exchange rate policy. Often, these policies have given rise to low and even negative real interest rates, thereby discouraging saving. Direct taxation of rural households, as through a land tax, was effective in transferring saving from agriculture in Japan during the first two decades of the 19th century (Kuznets, p. 47). However, direct taxation appears to be an unacceptable alternative to landowning classes, especially in light of the discrimination against agriculture that already exists in most developing countries.

II.A Interventions in Foreign Trade Markets

Governments of many developing countries rely on the use of trade and exchange rate policies to extract resources from agriculture and then to transfer them to the urban-industrial sector of the economy. Trade and exchange rate policies are employed, in part, because the instruments used to implement these policies are easily manipulated. Furthermore, they are subtle in the sense that their manipulation is not as directly observable by those from whom the resources are being extracted as are other forms of taxation, e.g., a land tax.

A recent USDA-University of Minnesota study provides insight into the control of foreign trade in food grains in a number of countries. Of the 21 countries investigated, all employed one or more of the following instruments to control foreign trade: an export-import state operated monopoly, import licenses, export taxes, or quotas. 19 countries reported the use of trade controls on wheat, 18 had trade controls on rice, 15 had trade controls on corn, 11 had trade controls on sorghum, and 4 had trade controls on millet. These instruments are most often used to maintain low and stable prices of food, relative to border market prices, by artificially raising imports above free-trade levels or by discouraging the export of food crops in which a country has a comparative advantage.

The experience of selected countries illustrates this point. Tanzania's National Milling Corporation is the agency with the primary responsibility for carrying out the country's stated policy of food grain self sufficiency. It maintains a statutory monopoly over the marketing and foreign trade in grains. Essentially, the agency enforces the government's domestic price controls by making the necessary adjustments in its stocks or imports in order to equilibrate domestic demand and supply at announced prices. This policy has caused substantial departures of domestic food grain prices from border market levels. The average ratio of domestic to border prices over the period 1964-1977 was 0.76 for maize, 0.64 for rice, and 1.15 for wheat.

The government's choice of domestic price levels was constrained by the nature of supply and demand for food and the competitiveness of the country's exports. The government's tendency to lower the price of maize relative to a less important crop in consumption was offset by the foreign exchange losses that would have been incurred for large departures between domestic and border prices of maize.

Efforts to transfer resources from agriculture often give rise to declining foreign exchange earnings and a crowding out of nonagricultural imports (Pitt). In the case of Tanzania, Lofchie points out that due to the severe shortage of foreign exchange and the urgent need to use remaining currency reserves to finance immediate food requirements, the government was compelled to impose stringent limitations on nonfood imports. These restrictions decreased the importation of economically important items such as raw materials for industry, new capital goods, and spare parts. The result was a serious economic depression. Hence, in some countries, efforts to transfer resources out of agriculture may actually limit resources available to the domestic industries the country is trying to protect.

This trade-off between the control of food imports and the scarcity of foreign exchange is less pronounced for industrial crops such as cotton, sugar, and coffee. Using similar policy instruments, these crops are often taxed to an even greater extent than are food crops. Bale and Lutz report domestic to border price ratios of 0.34 and 0.58 for cotton in Egypt and Pakistan, respectively. In Egypt during the late 1970's and early 1980's, domestic long-staple cotton prices were so depressed that domestic cotton-processing companies were forced to import cotton at border prices (i.e., at almost three times the price paid to domestic producers) in order to operate their plants at desirable capacities (USAID).

The transfers imposed on sugar and coffee producers in many countries are more complex because of the quota systems for these commodities. The U.S. quota price of sugar has been more than twice the world market price in recent years. Rather than permitting these rents to be captured by farmers, many sugar-exporting countries have imposed relatively high export taxes on sugar and many have chosen to produce a large share of national sugar production on state farms.

In the case of the Dominican Republic (Greene and Roe), export taxes on sugar in 1983 were about 36 percent of the fob export value. Moreover, the Consejo Estatal de Azucar (CEA), a Dominican state-owned enterprise, controls nearly 40 percent of the land planted to cane. The remaining revenues after taxes have been used to remunerate plants, equipment, and a large and growing state-employed labor force.

Policies to transfer resources from agriculture are made even more punitive to rural households when countries simultaneously employ measures to protect the domestic industrial sector.[5] Results from a recent IMF study of 35 developing countries find that the rates of protection of manufacturing are often higher than in most industrial countries. The average effective rate of protection was 50 percent during 1966-72 and 60 percent in the late 1970's (IMF 1985a, Table 64). Exceptions include countries such as Korea, Taiwan, Singapore, and Hong Kong. Many of the developing countries that have high rates of protection for manufacturing also allow imports of raw materials and intermediate inputs intended for export production to enter duty free (IMF, 1985a, p. 74).

Protection of the industrial sector directly affects agriculture in four important ways. First, rural households often face higher prices for agricultural inputs supplied by protected industries. Protection of import competing industries through tariffs or quotas restricts world market supplies from entering domestic markets at world market prices. Hence, domestic import-competing industrialists have little incentive to expand production beyond the demand of the domestic market. In this situation, domestically produced import substitutes are invariably produced at high unit cost. This often occurs either because the scale of the domestic market is sufficiently small to preclude the operation of plant and equipment at low unit costs or the licensing arrangements under trade protection are allocated to a limited number of firms on the basis of political patronage. Moreover, it is not unusual for the products of protected industries to be technically inferior to substitutes otherwise available in world markets. The result tends to be the production of inferior products by concentrated industries operating at high unit cost.

Second, either as a consequence of concentrated industries (and hence monopsonistic behavior), high unit costs, or both, agriculture often receives lower prices for commodities that undergo additional processing in the protected industries. In the case of Egypt, farm-level prices of commodities that underwent additional processing were often lower than the prices in the absence of protection. Egyptian cotton producers would have received higher prices if cotton had been exported rather than processed in Egyptian plants.

Third, protection makes the industrial sector appear more profitable than agriculture and, consequently, agriculture is forced to compete for resources that are artificially made more dear. This includes peak seasonal demand for labor and credit. Agriculture must also compete for public investment. If these investments do not take into consideration the artificially induced profitability of returns in the protected sectors of the economy, then public investment in agriculture is likely to be less than it would be in the absence of protection.

Fourth, returns in agriculture benefit from public investment in activities where markets function poorly. Hence, to the extent that interventions decrease the public sector's capacity to make these investments, agriculture and the economy must forego this potential source of gain in productivity.[6]

Trade interventions give rise to a number of direct and indirect effects that can alter the terms of trade between food and nonfood crops and between the agricultural and industrial sector. The prices of noninternationally traded commodities (mostly perishables such as fresh fruits, vegetables, and, depending on the country, livestock products) can also become distorted as resources flow out of commodities whose prices are distorted downward and into the production of these nontraded commodities. To the extent that the nontraded commodities are substituted in consumption for the commodities whose prices have been reduced, the demand for the noninternationally traded commodities tends to decline thus placing additional downward pressure on their prices.

The direct effects of trade interventions alone can lead to a transfer of resources from agriculture and to induce rural to urban migration. While the transfer of some resources from the sector is expected to occur in the process of growth, extensive interventions in trade artificially induce these transfers so that many of the other adjustments, if they occur at all, tend to occur at a reduced rate. These include the process of capital deepening and the development of infrastructure and other characteristics of growth and development in agriculture.

Before considering these in more detail, I turn to a second set of interventions, common in many developing countries, that often further exacerbate the problems faced by rural households.

II.B Production and Marketing Controls

Interventions in foreign trade markets that induce a transfer of resources from agriculture invariably lead to depressed conditions in the sector. Some governments react with policies that subsidize agricultural inputs and raise farm-level commodity prices while, at the same time, maintaining low and stable food prices to urban consumers.

These policies lead to a narrowing of the marketing margin and, in the extreme cases of Egypt (von Braun and de Haen, USAID) and Peru , to farm-level prices that are higher than their equivalent retail counterparts. Without subsidies, the narrowing of the margin implies lower returns to the resources employed in marketing activity and hence an exodus of merchants and middlemen traditionally involved in these activities. The implementation of the policy often amounts to the taking over of marketing functions by government agencies and state-owned enterprises. For many countries in Africa, these structures, in the form of marketing boards, have existed since colonial rule.

Some insight into the pervasiveness of production and product marketing controls can be obtained from the USDA-University of Minnesota study of food policies in developing countries. All 21 countries in the study were found to employ some type of domestic production and/or marketing controls for food grains. These included procurement, processing, storage, and transportation. These controls were implemented through licensing, subsidy schemes to middlemen, and, most commonly, through state-owned enterprises. The extent of control in a country tended to be in direct proportion to the expenditure share of the crop in household consumption. 18 countries imposed marketing controls on wheat, 19 imposed controls on rice, 13 imposed controls on maize, and 14 imposed controls on sorghum. The African countries in the study tended to employ the largest array of controls over the most crops, followed by Asian and then Latin American countries.

The direct budget expenditures from implementing a policy of buying dear and selling cheap are often increased by the losses that seem naturally to arise from the inefficiencies common to many state operated enterprises. In the Dominican Republic, the state-owned enterprises that displace private enterprise in agriculture include the Instituto Nacional de Estabilizacion de Precios (INESPRE) and the previously mentioned sugar enterprise, CEA.

INESPRE's statutory objectives are to regulate the prices of agricultural products in domestic markets and to protect consumption levels (IBRD 1985a, p. 34). Essentially, this agency is the counterpart of Tanzania's National Milling Authority. It buys and sells products at different points in the marketing chain and in international markets so that domestic markets clear at target prices. It also stores commodities to dampen the annual variation in market prices. In 1983, for example, INESPRE had accumulated stocks valued at one-third of an entire year's production of rice plus $14 million in stocks of maize, edible oils, and soybean meal (IBRD 1985a).

The extent of CEA and INESPRE's involvement in agriculture can be gleaned from their annual current operating budgets. Their combined average annual current expenditures amounted to nearly 40 percent of agriculture's GDP over the period 1976 to 1984. While these enterprises are known to contribute to the central government's budget deficit, estimates of their operating deficits are difficult to obtain. Conservative estimates of their average annual deficits are five percent of agriculture's GDP over the period 1976 to 1984.

Deficits incurred by state-owned enterprises can be large. Estimates of losses associated with state-owned enterprises of all types in seven countries ranged from a low of under one percent of GNP in Korea to over ten percent of GNP in Sri Lanka (Short). Their losses also appear to be an important factor explaining the need of some countries to restructure external debt. This is discussed in more detail in Section IV.

III. SOME MACRO-ECONOMIC EFFECTS OF INTERVENTIONS

The direct effects of interventions give rise to a host of indirect effects. The latter invariably involve an overvalued currency combined with implicit import subsidies, export taxes, and deficits on a country's trade account. These indirect effects cause additional distortions in the terms of trade within and between the agricultural and industrial sector, which, in turn, serve further to extract resources from the agricultural sector.

The magnitude of some of these distortions is indicated in recent studies of Egypt, the Dominican Republic, and the Philippines. Scobie found that central government budget deficits associated with Egypt's food subsidies were met by both foreign and domestic borrowing. The concomitant expansion of the money stock led to an excess supply of money balances and an excess demand for goods, both foreign and domestic. A 10 percent rise in government expenditures was found to increase inflation by about 5.3 percent, decrease the stock of net foreign assets by 1.7 percent, and devalue the Egyptian pound on the black market by about 3.3 percent.

Because food imports were the key to equilibrating domestic supply and demand at announced prices, a decline in foreign exchange was met first by postponing the import of capital goods and raw materials, which had deleterious effects on the output of industrial goods. The economy was thus made vulnerable to fluctuating world prices of food imports. A 10 percent deviation from trend in total industrial imports tended to decrease industrial output by 8.3 percent and investment industrial capital by about 8.8 percent. Taking into consideration the share of foreign exchange allocated to food imports, these estimates implied that a 10 percent increase in the price of imported food resulted in a drop in industrial output by 1 to 2 percent.

In the case of the Dominican Republic, the direct effects (nominal rates of protection) of interventions in foreign trade, procurement, and marketing on producer prices of sugar, coffee, and rice relative to a price index of industrial goods suggested that rice producers received a small implicit subsidy over the period 1966 to 1985 (with the exception of 1973 and 1974 when they received a fairly large implicit tax). Over the same period, sugar producers were implicitly taxed in most years and coffee producers were implicitly taxed in all years except four (Greene and Roe). Thus, the direct effects discriminated against the agricultural export crops, and to a much lesser extent against the main food crop (rice) relative to urban industrial goods.

However, estimates of the average annual overvaluation of the Dominican currency relative to the dollar from 1977 to 1984 ranged, in real terms, from 10 percent to 22 percent, depending on various estimates of implicit tariffs, taxes, and excess demand and supply elasticities.[8] In this case, the total direct and indirect effects of intervention on producer prices of sugar, coffee, and rice relative to a price index for industrial goods was estimated to average -33.1 percent, -38.0 percent, and -6.0 percent, respectively.

The traditional agricultural export crops were thus even more heavily taxed relative to producers of domestic industrial commodities. Rice producers were also taxed, albeit at a lower rate, relative to producers of domestic industrial commodities. Yet, the effect on rice production was significant. Estimates from an econometric model (Roe and Senauer) of the Dominican rice economy suggested that in the absence of distortions, rice production would have exceeded observed levels by an annual average of about 19 percent since 1980. Interventions also restrained the country's participation in foreign trade. Since 1977, exports averaged about 20 percent of real GDP. In the absence of interventions, it is estimated that exports would have averaged about 33 percent of real GDP (Roe and Greene).

In the case of the Philippines, Bautista also found that interventions since the 1950s consistently discriminated against agricultural export production in favor of home goods and import competing industries. He concludes, "Correcting the incentive bias against agricultural export production represents a potentially significant source of growth in agricultural income and foreign exchange earnings. Institutional changes, new technologies, infrastructure development, and other productivity-raising public investments may be necessary to boost significantly the long-term export performance of Philippine agriculture. However, they are likely to prove inadequate if relative incentives continue to be biased against agricultural export production."

As the evidence illustrates, interventions have

altered the course of economic development in many countries. Indirect effects have come about in part because interventions have contributed to increased government expenditures that have exceeded their fiscal capacity to meet these costs. Associated with these expenditures is an increase in a country's currency, an increase in inflation, a decrease in real interest rates[9], and an increase in the real exchange rate that serves further to increase the implicit subsidy to food imports and to tax exports.

The consequences of interventions can lead to a change in the domestic terms of trade against agriculture and in favor of the urban-industrial sector. Naturally, this leads to an undervaluation of agricultural resources, an outflow of capital from the agricultural sector, and an increase in rural to urban migration. The increase in urban population would seem to place additional pressure to lower the prices of wage goods, primarily food staples. These adjustments slow economic growth in agriculture and, therefore, agriculture's contribution to the growth process.[10] Furthermore, these interventions serve to alter a country's international terms of trade. In other words, the process of transferring resources from agriculture by means that give rise to the distortions discussed here invariably leads to a "withdrawal" of a country from international markets. The efficiency gains to domestic resources from economies of scale and specialization that world markets provide are reduced. In the longer run, these efficiency losses limit a country's capacity to supply goods and services to a growing population.

Consider, for example, the impact on rural households. In the process of economic growth, rural households can be viewed as undergoing a vertical disintegration--a movement toward specialization of production activities, with an increasing share of household expenditures on preferred foods, housing, clothing, and other nonfood items. Even in the presence of large productivity increases in agricultural output, income and population growth can increase the demand for food and feed grains and, in some countries, increase imports of both (Mellor and Johnston). As productivity increases, the opportunity cost of time to the household increases. Labor is allocated away from labor-intensive activities and more reliance is placed on the market for goods and services otherwise produced in the more traditional household.

For rural food-surplus households, the means used by many governments to transfer resources from agriculture clearly serve to retard this entire process because the returns to agricultural resources are artificially reduced. Rural labor-surplus food-deficit households are also adversely affected. While food prices may be lower than they might otherwise be in the absence of interventions, rural employment opportunities are reduced. The additional employment opportunities in urban-industrial areas created by the import substitution-industrialization policies are usually not sufficient to pull the surplus labor from agriculture that these policies have effectively displaced. While real wages may be higher in urban areas, the capital-to-labor component of the technology of the new industrial plants is often capital-intensive relative to a developing country's endowments.

Furthermore, the skill levels required of labor to operate these plants may, in any case, exceed the levels of rural labor. In many countries, population growth coupled with insufficient labor absorption by the industrial sector has resulted in a decline in the land-labor ratio and, in the absence of technical change and increased capital inputs, a decline in the real wage (Hayami and Ruttan, Table 13-1). Attempts to circumvent this problem by state-owned industrial enterprises seem only to exacerbate the problem.

Public investment in areas where markets perform poorly (rural infrastructure, agricultural research, rural education) serves to enhance market linkages with rural households. For example, investment in roads lower spatial costs and, thereby, the marketing margin between farm and wholesale-level markets. Effectively, this improves the terms of trade for market goods relative to home-produced goods, thereby accelerating the vertical disintegration of rural households. To the extent that interventions decrease public resources available for investment in these areas, rural income streams and the capital deepening process associated with productivity increases in agriculture are diminished.

IV. SUSCEPTIBILITY TO SHOCKS

Countries pursuing the types of policies considered above tend to be vulnerable to shocks to the world markets, such as those that occurred during 1973-1974 and again in 1979-1980.[11,12] Their vulnerability arises because governments are either reluctant to alter policy in light of shocks or they do so with a considerable lag. These policies become difficult to manage and maintain in turbulent world markets because low-cost food and import substitution-industrialization policies are, for the most part, dependent on interventions in the trade sector. Since the source of public revenue is primarily from taxes and tariffs on exports and imports, world market shocks that adversely alter a country's terms of trade also adversely affect a country's fiscal capacity to carry out programs and maintain subsidies without incurring fiscal imbalances, let alone trade imbalances.

It is evident from Scobie that Egypt's policies were a fundamental determinant of the level of capital flows, the efficiency with which capital was used (e.g., investment in productive activities compared to consumption subsidies), and the country's capacity to service its debts from export earnings. When food imports are required to equilibrate demand and supply at announced prices, shocks that adversely affect a country's terms of trade can, in the absence of other adjustments, increase food subsidies and the level of protection to otherwise noncompetitive industries. Consequently, unsustainable government budget deficits can occur. A frequent result is an increase in a country's domestic and foreign borrowing. This debt serves to increase aggregate demand and to further exacerbate the distortions discussed previously. Furthermore, interventions tend to prevent world market signals from being transmitted to the private sector, and resource adjustments that would otherwise take place either do not occur or do so with a considerable lag. In the absence of interventions, adverse shocks to a country's terms of trade would tend to decrease its imports and, through adjustments in capital markets, lower the country's standard of living relative to countries whose terms of trade have improved.

Another measure of a country's susceptibility to shocks in world markets is the effect they have on the probability that the country will need to restructure its foreign debt. In Chipman, et al., a probit model was fit to data on 17 countries for the period 1975 to 1983. Of the five explanatory variables, the two that explained the largest variation in the probability of restructuring was the World Bank's index of price distortion (IBRD 1985b, Table 4.1) and the ratio of debt held by state-owned enterprises to GNP, lagged two years. This model predicted 90 percent of the countries that rescheduled their debt in 1984. The countries with the largest index of price distortion and largest ratio of non-central public sector debt to GDP tended to be those that pursued extensive interventions of the type discussed here. These results suggest that many countries used debt to cover the fiscal imbalances due to interventions instead of making capital investments that earn a flow of returns to meet payments on principal and debt service. Rising real interest rates, appreciation of the dollar, and declining foreign exchange earnings, which in part was due to trade interventions, give rise to unsustained levels of foreign debt in many countries. The restructuring exercise generally requires the debtor country to undertake adjustment policies. In the short run, these policies seem to have led to considerable adjustment difficulties for low income households. These difficulties might have been avoided if the countries instead had chosen to liberalize their policy and reduce the level of government deficits over a longer period of time. Ironically, the recent decline in the value of the dollar and in real interest rates may allow some countries to avoid liberalization of their policies.

V. SOURCES OF RESISTANCE TO LIBERALIZATION

If the social economic losses from these policies are large, why in general are they pursued? While surely incomplete, three possibilities are considered: (1) policies are the outcome of political pressures exerted by members of the domestic economy seeking their own interests, (2) policies are mistakes, or more generally, failures of the planning process, and (3) policies, when first implemented, may have implied small social costs but, with the passage of time, they become difficult to change because the adjustment cost incurred by groups otherwise benefiting from the policy may be high, and these costs may be disproportionately borne by the poor. (I have omitted the arguments advanced by Prebisch and de Janvry. The reader is referred to Spraos and Bates, p. 166-169 for a review and short critique of the Prebisch thesis and to Schuh (1984) for a critique of de Janvry's argument.)

Some recent insights into the factors motivating government intervention can be found in Bates and Colander. Bates rejects the notion that governments intervene so as to secure the best interests of their societies. Essentially, he accepts the hypothesis stated in (1) above. He argues that this view is consistent with the observation that urban households are potent pressure groups demanding low-priced food. They are potent because they are geographically concentrated and strategically located. They can quickly organize and they control public services so that they can impose deprivation on others. Bates supports this observation by noting that urban unrest forms a significant prelude to changes of governments in Africa.

Interests of urban consumers coincide with those of domestic industrialists who view low-priced food as serving to decrease the pressure on wages. The industrialists also are effective in obtaining protection from imports because of the notion, common in many circles, that the key to development lies in industrialization. And, in any case, since industrial goods account for a small share of most households' budgets in LDC's, price discrimination in favor of these goods will not have a large negative impact on the welfare of most households. The result is policy which tends to support import substitution and, simultaneously, low-cost food to urban households.

The same argument applies to developed economies. In advanced stages of development, the food share of the budget declines so that consumers become less sensitive to increases in food prices. Agriculture is a smaller component in the total economy, and farmers are more specialized. Within their area of specialization, they are better able to organize then are urban groups. This situation is virtually the reverse of the case for developing countries. With food a small share of consumers' expenditures, protective demands in agriculture can be met at lower economic cost to urban households. The result is that the agricultural sector is likely to receive protection at the expense of the industrial sector.

Others also seem to support this general view (Hayami and Honma and Hayami, Anderson, 1983, 1985). They extend it to explain policy regimes in developed countries that protect agriculture and regimes in developing countries that tax agriculture. Anderson (1983) notes in his study of the growth of agricultural protectionism in East Asia that countries tend to switch from taxing to subsidizing agriculture in the course of economic development. And the timing of this switch is associated with agriculture's declining comparative advantage relative to manufacturing.

While these arguments provide insights into the

motivation for interventions, it is not clear why governments prefer to intervene in markets. To accomplish many of the same objectives, they could intervene in areas where markets function poorly. Bates (pp. 173-178) argues that market interventions facilitate the allocation of political rents. In his terminology, market interventions facilitate the "organization of the rural constituency" who support the government and to "disorganize the rural opposition." Markets fail in the provision of public goods because of the free rider problem. They fail largely for the same reason in the provision of political rents.

Because of the free rider problem, Olson argues that political coalitions are likely to be narrowly based and interested in the distribution of wealth rather than in attempts to allocate resources to increase society's output. Drawing on Olson and Buchanan, Srinivasan (1985) argues in the case of India that the policies that sought to alleviate the conditions of the poor were not undesirable per se. Instead, "It is that the policies introduced in the name of poverty alleviation increased the power of other rent-seeking distributional coalitions."[13] Market interventions tend to be more effective in capturing rents for these coalitions than interventions in areas where markets function poorly. Srinivasan (1985) adds that countries that follow inward-oriented development strategies of import substitution-industrialization are more prone to trigger these activities than are countries following outward-oriented strategies. It would seem that these arguments might also be extended to explain, in part, the formation of state-owned enterprises which permit the capturing of rents by directors and employees of the enterprise.

The relaxation of interventions that distort an economy confronts the political forces that have gained from the distortions. In the short run, it is possible that extreme shocks to an economy are required to dislodge the political structures that have given rise to costly forms of intervention. In the longer run, education and technical changes in agriculture that significantly alter income streams (such as the green revolution) also induce changes in institutions. Whether these changes can come about in highly distorted economies and, if they do, whether they will be sufficient to induce changes in policy is open to question.

Perhaps in all countries, some interventions and the manipulation of policy instruments are simply the result of policy mistakes. In practice, numerous government agencies are involved in the planning-policy implementation process. Most projects have spatial, temporal, and commodity target-group specificity. This process is complex, and characterized by a multiplicity of policy instruments and a maze of projects.

The development planning literature (e.g., Agarwala, Cochrane, and Stopler) has documented the experience of many countries where the mismanagement of this complex process and the development of plans based on faulty cause and effect and program-project implementation assumptions[14] have given rise to outcomes that bore little resemblance to initial intentions. Since physical and administrative infrastructures are poorly developed in many developing countries, it is difficult to target interventions (taxes, food subsidies) in ways that have minimal market distorting effects. Organizational problems often give rise to poly-archical decision making structures rather than hierarchical. That is, interventions are not always centrally directed. Instead, interventions are often carried out by semi-autonomous agencies and state-owned enterprises without the direct control or knowledge of a country's central planning-policy making authorities. While these factors are not independent of the political forces mentioned above, they must surely affect the patterns of interventions and the welfare gains and losses in many countries.

Interventions that have been in place for an extended period of time can induce structural changes in an economy. Put another way, the value of protection gets built into the value of sector-specific assets so that in the short-run, policy liberalization can have significant wealth effects. An example is industrial plants and equipment that process specific commodities or fabricate particular goods, which, in the absence of protection, lose part of their value, the loss being greater the more difficult it is to transfer the capital to other enterprises.

The human capital employed in these enterprises will also be displaced, losing seniority rights and perhaps needing to undergo retraining to obtain equivalent wage levels in other activities. Some of the displaced workers may enter the surplus labor pools of the lower skilled, thereby placing downward pressure on wages in these markets as well. The end result can be lower earnings to unskilled labor so that lower income households bear a disproportionate loss in income compared with households of higher skilled, though perhaps displaced, workers. In this environment, households that had not previously been significant participants in the political process may, in the light of possible changes in policy, oppose change because of the short-run adverse wealth effects even though, in the longer run, they may gain from the reversal of these policies.

Furthermore, significant changes in policy imply that households need to change their expectations regarding the source and levels of future income streams. When 70 percent of disposable household income is allocated to food, unskilled labor from urban households with limited ties to rural resources tend to bear the brunt of the readjustment process. The possibility of lower incomes and the uncertainty this implies almost surely adds to the political forces mentioned above that resist changes in policy.

VI. CONCLUDING COMMENTS

The removal of interventions that give rise to the type of distortions considered in this paper will, almost surely, require a comprehensive plan that deals with the sources of resistance discussed in the previous section. The period of time required to carry out the plan in an orderly manner will likely take longer the greater are the distortions and the more entrenched are the enterprises that owe their existence to, and implement, the interventions. Substantial effort will likely be required to convince those who face adjustment costs of the long-run social cost of continuing these policies.

Issues that a plan will need to address include: (1) building and redirecting government agencies to design programs and implement projects that are socially profitable in areas where markets function poorly in the allocation of resources, (2) developing equitable means of divesting public enterprises and, for natural monopolies, finding forms of organization that give rise to least cost operations and pricing behavior, (3) instituting alternative forms of public revenue generation other than unequal tariff and tax rates on imports and exports that give rise to distortions, and (4) formulating policies to ameliorate adjustment costs faced by low income households.

Low income is the fundamental cause of hunger and, as noted in section IV, low income households tend to face the major cost of adjustments to adverse changes in their environment. Income depends on the households' access to factors of production, including skills, that generate income streams sufficient to satisfy basic nutritional and health needs.

It is generally agreed that programs and projects to transfer income must be targeted because otherwise distortions come about. Target interventions include food stamps, fair-price shops, school lunch programs, public works projects, training programs, and, more specific to low income rural households, input subsidies, and extension-education programs. A self-targeting program is one that attracts the targeted group because of differences in its characteristics relative to higher income households. These characteristics include the low opportunity cost of time and the consumption of less preferred foods, such as cassava.

The difficulties of targeting include: (1) identifying low income households whose "basic" nutrition and health needs are not met, (2) the programs and projects that target urban households may not be to target rural households, and (3) targeting programs and projects to low income households often gives rise to high administrative costs relative to market interventions.

The choice of interventions to assist low income households will need to be based on a balance between targeted and "self targeted" interventions. The key lies in striking a balance in the choice among market and nonmarket interventions. This balance should seek to minimize the resource cost between the administration of nonmarket interventions and the efficiency loss from market interventions. Hence, it may be desirable to introduce distortions in the market for cassava by subsidizing its retail price while maintaining the farm level price at its undistorted level.

Fair price shops, food stamps, and the provision of public goods (e.g., water, health facilities) that are effective low cost means of targeting urban households may not be an effective low cost means for targeting rural households. Hence, targeted interventions for rural households will likely need to take a different form than for those in urban areas. In either the case of rural or urban households, Srinivasan (1983) cautions that "leakages" in targeted programs need to be a major concern. Leakages occur when ineligible individuals are included in target groups through fraud or bad program design. Worse still are programs where many of the targeted are excluded. Srinivasan cites a case in India were a program for input subsidies to low income farm households was captured by higher income groups, thereby exhausting program funds for the targeted group.

Footnotes

[1] See Hayami and Ruttan, Chapter 6: Sources of Agricultural Productivity Differences Among Countries. They refer to capital deepening as an increase in internal resources in agriculture such as investments in land improvements, livestock, the use of modern technical inputs (chemical and mechanical technology) and increases in human capital.

[2] Under assumptions that essentially preclude market failure, and provided that lump sum income transfers are feasible, welfare economics suggests that a noninterventionist strategy can, in principle, maximize efficiency in exchange, production and overall efficiency. Within this context, Buchanan (p. 14) draws the implication that "So long as governmental action is restricted largely, if not entirely, to protecting individual rights, person and property, and enforcing voluntarily negotiated private contracts, the market process dominates economic behavior and ensures that any economic rents that appear will be dissipated by the force for competitive entry." The problem, of course, is that in developing countries lump sum transfers are not feasible and the conditions that give rise to market failure are thought to be common. These include: imperfect competition, externalities, public goods, and risk and information asymmetries (commonly referred to as moral hazard and adverse selection). Whenever these conditions prevail, collective action by producers or consumers or by government can, in theory, give rise to an increase in welfare without making any other member of the economy worse off. See Stiglitz for a general discussion of these issues.

[3] See also Bale and Duncan and Part III. "Distortion of Incentives" in Distortion of Incentives," T. W. Schultz, editor, Indiana University Press, 1978, for more discussion of this issue.

[4] Total expenditure elasticity with respect to a household's consumption of various goods and services has,

for our purposes here, the same interpretation as an income elasticity. The former is often reported in empirical studies because of the difficulty of measuring total household income and of obtaining unbiased estimates of total income elasticities.

5 For more insights into the economics of import substitution industrialization policies, see "Comparative Advantage and Development Policy Twenty Years Later" in Essays in Honor of Hollis B. Chenery, M. Syrquin, L. Taylor and W. Westphal, editors, Academic Press, New York, 1984, and J.N. Bhagwati, R. Brecher, and T.N. Srinivasan, "DUP Activities and Economics Theory," in NeoClassical Political Economy, D. Colander, editor, Cambridge: Ballanger Pub. Co., 1984.

6 There is some evidence to suggest that protection of the domestic industrial sector through trade interventions also adversely affect the production and transfer of agricultural technology by the private sector in some countries (Pray).

7 In Peru, ECASA purchased domestic rice at prices 30 percent above prices charged to consumers in 1982 (Orden, et al.).

8 The "official" rate was estimated as the weighted average of the parallel market rate and the rate offered by the central bank. The formula used to estimate the local currency to dollar exchange rate that might prevail in the absence of distortions in each period was:

$$E = \{B(Z)(P_m)^{1+\eta}(1+t_\eta)^{\eta}/A(W)(P_x)^{1+\varepsilon}(1-t_x)^{\varepsilon}\}^{1/(\varepsilon-\eta)}$$

where B(Z) and A(W) are functions of exogenous variables Z, W appearing in the aggregate excess demand and supply functions for imports (m) and exports (x); P_m and P_x denote an index of border prices, in dollars for imports and exports; t_m and t_x denote implicit net taxes on imports and exports; and e and n denote the aggregate price elasticities of excess demand and supply respectively. See Roe and Greene for a derivation of this formula.

9 Negative real interest rates arise in many developing countries because nominal rates remain fixed during periods of high inflation (IMF 1985b). At various times

during the 1970s, negative real interest rates were particularly severe in Brazil, Ghana, Jamaica, Nigeria, Peru, and Turkey (IBRD 1983, p. 58).

[10] Kuznets lists the potential contributions of agriculture as (1) the low cost supply of food and raw materials for processing, (2) a market for producer and consumer goods produced by domestic industry, (30 a source of factor contributions (labor, capital) to the industrial sector and (4) a source of foreign exchange earnings and a source of foreign exchange savings through the production of import competing products.

[11] Evidence compiled by the World Bank (1985b) tends to support this view over a large number of developing countries. In comparing the adjustment policies of inward oriented countries, Balassa (1984, 1985) found that these countries lost export market share and, not withstanding substantial foreign borrowing, grew less rapidly than outward oriented economies. The latter, relying more heavily on market forces, were found to adjust sooner to changing conditions in world markets by accepting a slowdown in economic growth while, at the same time, pursuing output oriented policies of export promotion. Over the entire period, they grew much more rapidly. Inward oriented countries were Egypt, Morocco, Philippines, Jamaica, Peru, Tanzania, Indonesia, and Nigeria. Outward oriented countries were Tunisia, Kenya, Thailand and the Ivory Coast.

[12] Briefly, the 1973/74 shock was characterized by fluctuations in prices of primary commodities, rising prices of energy products, and a slowdown in economic activity in the developed countries. The 1979/80 disturbance was characterized by another increase in energy prices, sharp increases in real interest rates, declining volume and declining terms of trade for commodity exporters (IMF 1984, 1985b).

[13] Since Anne Krueger's pioneering article on the political economy of rent seeking, it has become more evident that the process of seeking to distort incentives can induce an additional source of welfare loss. The core of the argument is that groups affected by interventions may engage in lobbying activities which consume resources that would otherwise be employed in productive activities.

[14] This situation occurs when the government's policy-decisionmaking apparatus designs and implements policies based on a false perception of the problem confronting the economy or the result that the manipulation of a policy instrument is expected to have. An example of the latter is the use of an import quota to protect the import competing sector of the economy from foreign competition based on the belief that a point will be reached where the protected sector can compete in world markets. The experience from numerous countries suggests that the protected sector seldom reaches a point where it can compete in world markets.

References

Agarwala, Romgopal. "Planning in Developing Countries: Lessons of Experience." IBRD Staff Working Papers No. 576, 1983.

Agency for International Development. "Agricultural Prices and Policies in Egypt (1974-1981)." Country Development Strategy Statement, February 1982.

Anderson, Kym. Economic Growth and Distorted Agricultural Incentives. Department of Economics, University of Adelaide, Adelaide, S.A., February 1985.

_________. "Growth of Agricultural Protection in East Asia." Food Policy, Butterworth and Co. Pub., November 1983:327-336.

Balassa, Bela. "Adjusting to External Shocks: The Newly Industrialized Developing Countries in 1974-76 and 1979-81." Discussion Paper No. DRD89, IBRD, Washington, D.C., May 1984.

_________. "Policy Responses to Exogenous Shocks in Developing Countries." American Economic Review, 76(1986):75-83.

Bale, Malcolm D. and Ronald C. Duncan. "Food Prospects in the Developing Countries: A Qualified Optimistic View." American Economic Review, 73(1984):244-248.

_________ and Ernst Lutz. "Price Distortions in Agriculture and Their Effects: An International Comparison." American Journal of Agricultural Economics, 63(1981):8-22.

Bates, Robert H. "Governments and Agricultural Markets in Africa" in Johnson, D. Gale and G. Edward Schuh (eds.) The Role of Markets in the World Food Economy. Westview Press, 1983.

Bautista, Romeo M. "Effects of Trade and Exchange Rate Policies on Export Production Incentives in Agriculture: The Philippines." Working paper, IFPRI, 1985.

Buchanan, J. M. "Rent Seeking and Profit Seeking" in J. M. Buchanan, R. D. Tollison and G. Tullock, (eds.) Toward a Theory of Rent Seeking Society. Texas A&M University Press, 1980.

Chipman, John, R. Holt, J. Turner and T. Roe. "Korea: An Economic and Political Analysis." The University Research Consortium, University of Minnesota, May 1984.

Cochrane, W. W. Agricultural Development Planning: Economic Concepts, Administrative Procedures and Political Process, Praeger, 1974.

Colander, D., ed. Neoclassical Political Economy. Cambridge: Ballanger Pub. Co., 1984.

de Janvry, Alain. "The Political Economy of Rural Development in Latin America" in Eicher, Carl and John M. Staatz, (eds.) Agricultural Development in the Third World. Johns Hopkins University Press, 1984.

Downs, A. An Economic Theory of Democracy. New York: Harper and Row, 1957.

Gerrard, Christopher D. and Terry Roe. "Government Intervention in Food Grain Markets: An Econometric Study of Tanzania." Journal of Develop. Econ., 12(1983):109-132.

Greene, Duty D. and Terry Roe. "A Comparative Study of the Political Economic of Agricultural Pricing Policies." Report prepared for the World Bank, January 1986.

Honma, M. and Y. Hayami. "Structure of Agricultural Protection in Industrial Countries." Faculty of Economics, Tokyo Metropolitan University, Mimeo, September 1984.

Hayami, Yujiro and Vernon W. Ruttan. Agricultural Development: An International Perspective. 2nd ed. New York: Johns Hopkins University Press, 1985.

International Monetary Fund. "Trade Policy Issues and Developments." Occasional paper 38, Washington, D.C., July 1985.

__________. "World Economic Outlook." Washington, D.C., 1984, 1985.

Krueger, Anne. "The Political Economy of the Rent-Seeking Society." American Economic Review, 64(1974):291-303.

Kuznets, S. "Economic Growth and the Contribution of Agriculture" in Eicher, C. K. and L. Witt (eds.) Agriculture in Economic Development. New York: McGraw-Hill, 1964.

Lofchie, M. F. "Agrarian Crisis and Economic Liberalization in Tanzania." Journal of Modern African Studies, 16(1978):451-475.

Mellor, John W. and Bruce F. Johnson. "The World Food Equation: Interrelations Among Development, Employment, and Food Consumption." Journal of Economic Literature, XXII(1984):531-571.

Olson, M. The Logic of Collective Action. Cambridge: Harvard University Press, 1965.

Orden, David, Duty Greene, Terry Roe, and G. Edward Schuh. Policies Affecting the Food and Agricultural Sector in Peru, 1970-1982: An Evaluation and Recommendations. Report prepared for USAID, Department of Agricultural and Applied Economics, December 1982.

Pitt, Mark M. "Alternative Trade Strategies and Employment: Indonesia" in A. O. Krueger, H. B. Lary, T. Monson, and N. Akasanee (eds.) Trade and Development in Developing Countries, Vol. 1. Chicago University Press, 1981.

Pray, Carl. Private Sector Research and Technology Transfer in Asian Agriculture: Report on Phase I AID Grant OTR-0091-G-SS-4195-00. Bulletin No. 85-5, Economic Development Center, Department of Economics and Department of Agricultural and Applied Economics, University of Minnesota, December 1985.

Prebisch, R. "Commercial Policies in Underdeveloped Countries." American Economic Review, 49(1959):251-273.

Roe, Terry and Ben Sanauer. "Simulating Alternative Food-Grain Price and Trade Policies: An Application to the Dominican Republic." Journal of Policy Modeling, 7(1985):635-648.

_________ and Duty Greene. "The Estimation of a Shadow Equilibrium Exchange Rate." Working Paper, Department of Agricultural and Applied Economics, University of Minnesota, March 1986.

Schuh, G. Edward. Food Aid as a Component of General Economic and Development Policy, paper presented at conference on Improving the Developmental Effectiveness of Food Aid in Africa, sponsored by Ag. Development Council, Inc., Abidjaw, Ivory Coast, August 24-26, 1981.

_________. "The Political Economy of Rural Development in Latin America: Comment" in Eicher, Carl K. and John M. Staatz (eds.) Agricultural Development in the Third World, Johns Hopkins University Press, 1984.

Schultz, T. W., ed. Distortion of Agricultural Incentives. Bloomington: Indiana University Press, 1978.

Scobie, Grant M. "Food Subsidies in Egypt: Their Import on Foreign Exchange and Trade." Research Report 40, IFPRI. August 1983.

Short, Peter. Appraising the Role of Public Enterprises: An International Comparison. IMF Occasional Paper Series, Washington, D.C., 1983.

Spraos, John. Inequalising Trade: A Study of Traditional North/South Specilisation in the Context of Terms of Trade Concepts. Oxford University Press, 1983.

Srinivasan, T. N. "Hunger: Defining It, Estimating Its Global Incidence, and Alleviating It" in Johnson, D. Gale and G. Edward Schuh (eds.) The Role of Markets in the World Food Economy, Westview Press, 1983.

_________. "Neoclassical Political Economy, the State and Economic Development." Asian Development Review, 3(1985):38-58.

Stiglitz, Joseph E. "Some Theoretical Aspects of Agricultural Policies." Research Observer, Vol. 2, No. 1(1987):43-60.

Stopler, Wolfgang. Planning Without Facts. Harvard University Press, 1959.

Timmer, C. Peter, Walter P. Falcon and S. R. Pearson. Food Policy Analysis. Johns Hopkins University Press, 1983.

U.S. Department of Agriculture. Food Policies in Developing Countries. Foreign Agricultural Economic Report Number 194, December 1983.

Von Braun, Joachim and Hartwig de Haen. "The Effects of Food Price and Subsidy Policies on Egyptian Agriculture." Research Report 42, IFPRI, November 1983.

World Bank. "Dominican Republic: Economic Prospects and Policies to Renew Growth." A World Bank Country Study, Washington, D.C., 1985a.

_________. "The World Development Report 1985." Washington, D.C., 1985b.

_________. "Poverty and Hunger: Issues and Options for Food Security in Developing Countries." A World Bank Policy Study, Washington, D.C., 1986.

Table 1. Total Expenditure Elasticities and Expenditure Shares for Food, Housing, Clothing, and Other Nonfood Items, 4,028 Households, Dominican Republic.

Exp. Group[1]	Mean[2] Exp.	Mean[2] Income	Food Total[3] Exp. Elast.	Food Share Exp.	Housing Total Exp. Elast.	Housing Share Exp.	Clothing & Other Total Exp. Elast.	Clothing & Other Share Exp.
RLY	71.9	82.9	1.21	0.73	0.66	0.21	0.01	0.07
RMY	128.7	139.5	1.26	0.71	0.63	0.17	0.01	0.12
RHY	264.7	304.8	0.68	0.63	0.86	0.17	1.53	0.20
ULY	111.6	120.8	0.91	0.64	1.04	0.25	1.37	0.11
UMY	223.6	250.4	0.85	0.55	1.28	0.24	1.07	0.21
UHY	610.3	646.1	0.32	0.39	0.26	0.30	2.21	0.32

Source: T. Yen, Stagewise Estimation of a Complete Demand Systems with Limited Dependent Variables and Nonlinear Constraints: An Application to Dominican Household Consumption Data, Unpublished Ph.D. Thesis, Department of Agricultural and Applied Economics, University of Minnesota, St. Paul, April 1986.

[1]RLY denotes the category of rural low income households, ULY denotes urban low income households, etc. The expenditure categories are based on monthly incomes as follows: for rural households: RLY < \$100; \$100 < RMY < \$160; \$160 < RHY, and for urban households: ULY < \$165; \$160 < UMY < \$300; and \$300 < UHY. The official rate of Peso-Dollar exchange at the time this data was collected was 1:1.

[2]Expenditures are mean total monthly expenditures on all goods and services. Income is mean monthly income including income in kind. The residual between expenditures and income includes savings and other unaccounted for income.

[3]The first column denotes total expenditure elasticities with respect to the consumption of food, housing, and clothing, and other goods and services. The second column denotes the share of total expenditures allocated to the particular category (e.g., food) listed at the top of the column. See footnote 4 for further details.

6

The (Negative) Role of Food Aid and Technical Assistance

Melvyn Krauss

Few economists argue that foreign aid can substitute for sound economic policies in recipient countries. What many do maintain, however, is that aid is better than nothing--that it can provide relief for people who suffer from extreme poverty. This argument, however, is misleading for several reasons. First, economic aid may not reach the poor people for whom it is intended. Just because economic aid is given for the benefit of poor people does not mean that the poor actually receive it.

"Foreign aid," writes Peter Bauer, "does not in fact go to the pitiable figures we see on aid posters, in aid advertisements, and in other aid propaganda in the media. It goes to the governments, that is to the rulers, and the policies of the rulers who receive aid are sometimes directly responsible for conditions such as those depicted. But even in less extreme instances, it is still the case that such aid goes to the rulers; and their policies, including the pattern of public spending, are determined by their own personal and political interests, among which the position of the poorest has very low priority."[1]

A 1981 story in The Wall Street Journal confirms that rather than go to poor people in real need, food aid often winds up being used by the governments of poor countries to subsidize elites and keep themselves in power:

> No wonder the effectiveness of food aid around the world is under suspicion in the United States. Some critics have concluded that this low-cost food is merely a device for keeping elites in power by

> propping up foreign-government budgets and feeding influential middle classes. Food aid has discouraged food production, these analysts say, and has failed to address the basic challenge of helping the poor earn enough to buy the food already available.[2]

In Bangladesh, for example, food aid meant for starving people never reaches them because the government used the food to buy votes. The Wall Street Journal story continues:

> It comes as a surprise to a layman, but not at all to the experts that food aid arriving in Bangladesh and many other places isn't used to feed the poor. Governments typically sell the food on local markets and use the proceeds however they choose. Here, the government chooses to sell the food in cut-rate ration shops to members of the middle class.[3]

In these shops the food is sold at extra-low prices, so that ration cards are need to equate demand and supply. The key to getting subsidized food, therefore, is a ration card, which is distributed to elites both inside and outside the government. The inequity of food aid that cannot go to the poor people for whom it was intended because the poor can't get ration cards is obvious. So is the blatant political use of food aid. Bangladesh Food Minister Abdul Momen Khan is quoted: "To be very frank, the political aspects (of dismantling the ration system) must be taken into consideration." In other words, if the starving people actually did get the food aid intended for them, the corrupt government, responsible for their poverty, could no longer maintain itself in power by buying support with subsidized food.

The Bangladesh case of misused food aid is by no means an isolated one. Indeed, compared to the recent tragedy in Ethiopia, it is a distinctly minor one.

Many of us no doubt remember the heart-breaking pictures and television reportage of starving adults and children in that poverty-stricken North African country. These people were dying a most horrible death before our very eyes--indeed, as we sat in the comfort and security of our living rooms watching their cruel fate on television. Who was not moved to dig into their pockets to give, and give generously, to famine relief in the face of such calamity?

It goes without saying that all three U.S. television networks--CBS, NBC, and ABC--assured their viewers at the time that the deaths they were viewing on the screens were due to famine--a natural not man-made calamity. But according to a group of concerned physicians, Doctors Without Borders, "more Ethiopians were dying as a result of the policies of their Government than as a result of famine." Their report, entitled "Mass Deportations in Ethiopia," says that as many as 300,000 people are likely to die in the forced re-location from North to South, a death rate of 20 percent. The information in the report came largely from relief workers and Ethiopian refugees in the Sudan.

According to a story in The New York Times, "the doctors" report says that thousands of people have been resettled 'at gunpoint', that families have been separated, that food and blankets from abroad have been 'used as bait', and that conditions during transport and in the resettlement areas have led to widespread disease and death."[4]

While Ethiopian officials claim that resettlement was necessary because the dense population of Ethiopia's northern plateaus had exhausted the land, informed sources contend that the true purpose of the program is to depopulate the northern areas where anti-Marxist rebel groups receive support from the masses and set up Government-dependent colonies and Soviet-style collective farms in the new areas of settlement.

Doctors Without Borders calls the re-location program "one of the most massive violations of human rights we have seen" and notes that it is "being carried out with funds and gifts from international aid."

Some days after Doctors Without Borders released their report, The Wall Street Journal published the following editorial: "French relief workers were touring Washington last week with awful news that had already been reported by eyewitnesses from the U.S. Agency for International Development. The Ethiopian government's year-old 'relocation program', now greatly stepped up by the military Dergue, or junta, and its Russian patrons, has already claimed the lives of 20 percent of its targets, a death toll of some 100,000. It shapes up as a mass extermination in the order of the Khmer Rouge killing fields and the deportation of Armenians in 1915, with the added horror that it would not have been possible without the aid and silence of Western famine relief."

"Famine relief trucks have been diverted to move people, while grain rots at the ports. The roundups have disrupted harvests and forced abandonment of whole herds of livestock. Grain has been taken from famine areas and sent south to maintain concentration camps. In the meanwhile, government troops have launched their biggest offensives ever in the heart of the famine regions, drawing logistical support from the relief stockpiles while burning the rebels' crops."[5]

Unhappily, the Ethiopian case is not the only one where food aid financed malevolent government persecution of innocent peasants. According to Peter Bauer, "it has been widely and rightly recognized, even by supporters of President Nyerere [of Tanzania], that without large-scale external aid he would not have been able to persist for so many years with forced collectivization and large-scale removal of people into so-called socialist villages. These policies not only involved brutality and hardship but had devastating effects on food production and distribution, and on economic conditions generally."[6]

The Ethiopian case shows that food aid helps bad governments do bad things to their people. Such "bad things", of course, are not always as extreme as Nazi-like forced deportations. But food aid can--and often does--facilitate classic policy errors in recipient countries, which make already poor countries even poorer.

A specific example where food aid facilitated bad general economic policy in India, where the strategy of its second and third 5-year plans focused on developing import-substituting industry, particularly heavy industry.

It is generally agreed that India could not have pursued misconceived strategy, which transfers domestic resources from high to low productivity uses, had it not been for generous food aid from abroad, particularly from the United States. Even friends of foreign aid, such as liberal economists Paul Isenman and Hans Singer, point the finger of blame at food aid in this case. They wrote, "food aid supported and facilitated the [import-substitution] strategy, primarily by enabling the Indiana government to maintain large subsidized distribution programs while, in the eyes of most analysts, not adequately addressing some basic questions of food grain production and distribution."[7]

The point is that import-substitution artificially transfers resources out of agriculture into industry. This reduces agricultural production--which raises the price of food unless food can be imported from abroad. Note that without such food transfers, import substitution would be compelled to cease, particularly in a democratic society like India where government could be expected to be responsive to popular discontent over food shortages and high food prices. This example from India illustrates what may be called the "food aid trap." The demand for food aid in poor Third World countries indicates severe economic mismanagement. Yet, satisfying that demand only serves to perpetrate the economic mismanagement responsible for the poverty in the first place.

So far, attention has been focused on the role food aid plays in facilitating bad or foolish policies by bad or foolish governments. Next, I would like to turn to the so-called "disincentive effects" of food aid--the direct role that food aid plays in damaging local agriculture in recipient countries.

By increasing the amount of food available for domestic consumption, food aid depresses the price of agricultural products in recipient countries. Such price reductions in turn depress agricultural output, so that what the recipient country gains in food aid it loses to some extent in domestic output. When the offset in domestic output equals the food aid, the marginal effect of the food aid on food availability is zero. In this case, the food aid totally fails to increase the recipient's domestic availability of food. The marginal effect is greater than zero but less than unity when the offset of domestic supply is less than the food aid. And it is even theoretically possible for the domestic supply offset to be greater than the food aid, in which case the marginal effect of the food aid is negative.

Whatever, the marginal effect of food aid on domestic food consumption, however, it is clear that it can have a negative effect--and sometimes a devastating negative effect--on domestic food production. For example, a study on food aid to Columbia, by Dudley and Sandilands, showed that, from 1953 to 1971, wheat imported under food aid rose from 50 percent to about 90 percent of total wheat consumption. In that same period, the price for wheat declined by about half, while wheat production declined very sharply to about one-third of its original level.[8] Note the direction of causality. It was not that there was food aid because of low food production, but that there was low food production because of food aid.

Moving down the supply curve is not the only consequence for oputut of price changes induced by food aid. There are also the income transfer effects of price changes to consider--and these too tend to reduce agricultural production in aid-receiving countries. Income transfer effects relate to the changes in the distribution of income caused by price changes. For example, when the price of agricultural commodities decline because of food aid, income is redistributed from local producers to local consumers. Whatever one's subjective evaluation of the ethics of such redistribution, it is likely to reduce agricultural production by reducing the means by which investments inn agriculture, either to extend the margin of cultivation or improve agricultural productivity through technological progress, can be made. When agricultural investment suffers, so must agricultural production.

An important consequence of food aid, then, is to reduce agricultural production capabilities in the long-run and make recipient countries more dependent upon foreign nations for their food supply than they otherwise would be. This criticism is not one of agricultural imports, _per se_--only _artificially_-induced increases in agricultural imports due to food aid. Such reduction in imports impose an unwarranted dependency of recipient countries on outsiders that can be justified neither by economic nor political considerations.

If food aid is so damaging to recipient countries, why then, do we have so much of it? The answer is simple. From the point of view of the recipient governments, food aid increases the resources it has at its disposal to effectuate its programs and maintain itself in power. In particular, food aid has facilitated a government-directed and largely unsuccessful process of industrialization through import substitution in third-world countries--at the expense of agriculture. Writes Dale E. Hathaway of the Ford Foundation:

> Despite their difficulties in expanding agricultural output, the under-developed countries have sacrificed the price incentive needed to speed the adoption of new technology to pursue a cheap food policy for their urban consumers.[9]

Matching the demand for food aid in underdeveloped countries has been a supply of food aid from the developed countries, primarily because food aid has become an integral part of the process of subsidizing farmers in the developed countries. Food subsidies in rich countries create surpluses, which somehow must be disposed of if they are not to become a political embarassment. Isenman and Singer write: "The political importance of farmers, who benefit from the increased demand caused by food aid, adds to the political attractiveness of food aid.[10]

Food aid, thus, is a zero-sum policy in which farmers in developed countries gain at the expense of farmers in less-developed countries--all under the pretense of feeding the destitute, avoiding starvation and improving the nutritional standards of the world's poorer peoples. It is a policy of selfish interest wrapped in humanitarian interest, damaging the very people it pretends to help.

Can the same harsh judgment made of food aid be made of technical aid as well? Certainly, in my own field of economics, the record of technical aid has not been very good. In a classic article on the subject, "Why Visiting Economists Fail," Dudley Seers writes:

> A large and growing number of economics are working for foreign governments as direct employees, as invited visitors, or more commonly as experts supplied under national or international technical assistance programs. It would be insidious to cite examples, but there is little doubt that many of us have been in some degree, failures. In the ministries of almost any underdeveloped country you can find cupboards full of the reports of economists, reports which have been talked about for a week or two, then put away and forgotten. Scores of "missions" and individual advisers are now overseas writing reports, many of which, it can safely be predicted, will suffer the same fate.[11]

Technical assistance, of course, not only consists of sending Western experts to the Third-World but in sending Third-Worlders to the West to study in the West's institutions of higher learning. Unhappily, the education received by Third-World students in the West, more often than not, is totally irrelevant to Third-World concerns. This is true in economics--where bright Third- Worlders often study rarified problems of economic theory--and, according to E. G. Vallianotos, a noted food expert, it also is true in agricultural science. According to Professor Vallianotos, "Foreign-education underdeveloped country agricultural scientists prefer to work on problems that have very little to do with the conditions of their agricultural economies and especially the small-scale, household rural needs of the peasants. It is no accident then that the green revolution elite established such cordial and lasting relationships with those who had the same education and had little if any sympathy toward the peasant. This intellectual and economic congeniality inevitably led the two elites to develop technologies that slowly turned out to be more and more forbidding to the rural poor."

Like food aid, technical aid, particularly in agriculture, tends to benefit rich elites rather than the poor people who need help most. Many point to the Green Revolution, sponsored by the Rockefeller and Ford Foundations, as a successful example of technical aid. But Professor Vallianotos writes:

> The seeds of the green revolution have for the most part not helped the poor farmers. This disparity in beneficial participation was first documented in Mexico, where the green revolution was born. It was the affluent farmer who benefited most from the agricultural research of the Rockefeller Foundation in Mexico. But this research has had little impact upon the practices of the hundreds of thousands of small, poorly educated corn farmers who make out their living from the poorer soils with undependable sources of water. The two most basic reasons that made the new technology readily available to the larger growers reflect, first, political decisions to favor the rich and, second, demand of the new crop varieties for fertilizer and plentiful and controlled supplies of water. The Mexican experience of making the rich grow richer was repeated in Morocco. More than 80 percent of Morocco's farmers were simply left out of the benefits of the new food technology.[12]

Just as the analogy between technical and food aid holds up with respect to aiding elites in the Third World, it also holds up with respect to aiding recipient governments carry out bad or foolish economic policies. Indeed, in general, it can be said that technical aid has a distinct bias against the market system and in favor of government.

An example of how technical assistance provided by the World Bank "crowds out" the private sector in less-developed countries is a recent $40 million loan to Kenya to help that African nation develop its geothermal energy resources. Along with the loan came consultant services, training, and studies to determine Kenya's geothermal potential. Now, in the United States we have some very efficient private geothermal companies like Magma Power in California. Government plays no direct role in the exploitation of this country's geothermal resources. One would think that what's good for the goose should also be good for the gander. Yet American taxpayers, through their contributions to the World Bank, are facilitating the "crowding out" of the private sector in Kenya by subsidizing the public sector--something we wouldn't dream of doing in our own country. This "crowding out" argument is one of the major criticisms of technical assistance.

Not all government activities are bad or anti-economic, of course, and to the extent that technical aid assists Third-World governments in legitimate activities, it can play a positive role in economic development. Infrastructure projects like public roads, port facilities, etc., which genuinely raise the productivity of factors of production in the private sector, are an example of a legitimate public activity. Note, however, that the infrastructure must raise private productivity, and that such increase in private productivity must be greater than the cost of the funds for the infrastructure to benefit the economy. Infrastructure merely for infrastructure's sake is a waste of money--and to the extent that technical assistance facilitates such waste, it is to be decried.

On the other hand, technical aid that prolongs life, such as recent advances made in vaccine research against malaria, can be justified not only on humanitarian grounds but as investments in human capital as well. Even if such aid is intermediated by government, it is to be applauded, bearing in mind, though, that the credit accruing to the rulers of the beneficiaries may be used by them to prolong their immiseration of the general populace. The trade-off between saving lives on the one hand, and prolonging a bad government in power on the other, must come down on the side of saving lives.

In conclusion, technical aid must be seen as part of a process of the international transmissions of information from the West to the Third-World. It is a fair statement that, in general, such transmission has been biased against the market place and for government direction of economic activity. "What the Third World learns from the West, or about it," writes Peter Bauer, "or about present and past economic relations between the West and Third World countries, comes from or is filtered through opponents of the market. They dominate international reporting, the wire services, documentary films, and entertainment. The academic contacts between the West and the Third World are also dominated by opponents of the market. This influence has been paramount in augmenting the financial and intellectual resources and in enhancing the prestige and effectiveness of Third World opponents of the market. The latter have a virtual monopoly in the local commissions and delegations of the United Nations and its specialized agencies and affiliates, in the research organizations financed by these organizations, in the local operations of the large Western foundations, in international academic exchanges, in the planning teams supplied to Third World governments by American and British universities.... Visiting academics address a public soaked in anti-market ideas, derived from practically all Western visitors and the textbooks and other economic literature reaching those countries."[13]

Footnotes

1) P. T. Bauer, Reality and Rhetoric: Studies in the Economics of Development (Cambridge, Mass.: Harvard University Press, 1984), pp. 49-50.
2) Barry Newman, "Bangladesh Provides Plenty of Ammunition for Critics of Food Aid," The Wall Street Journal (April 16, 1981), p. 1.
3) op. cit.
4) Clifford D. May, "Moving Ethiopians Causes a Dispute," The New York Times (January 27, 1986), p. 26.
5) The Wall Street Journal (January 27, 1986), p. 26.
6) P. T. Bauer, op. cit., p. 52.
7) Paul J. Isenman and H. W. Singer, "Food Aid: Disincentive Effects and their Policy Implications," Economic Development and Cultural Change, pp. 205-237.
8) L. Dudley and R. J. Sandilands, "The Side Effects of Foreign Aid: The Case of PL 480 in Columbia," Economic Development and Cultural Change, 23 (January 1975), pp. 325-336.
9) Quoted in E. G. Vallianatos, Fear in the Countryside (Cambridge, Mass.: Ballinger Publishing Co., 1976), p. 71.
10) Isenman and Singer, op. cit., p. 209.
11) Dudley Seers, "Why Visiting Economists Fail," Journal of Political Economy (August 1962), p. 325.
12) Vallianatos, op. cit., p. 56.
13) P. T. Bauer, "Hostility to the Market in Less-Developed Countries," in The First World and the Third World edited by Karl Brunner (Rochester, NY: University of Rochester Policy Center Publications, 1978), pp. 172-73.

7

"The (Negative) Role of Food Aid and Technical Assistance": A Comment

Don Paarlberg

Mr. Krauss presents a sweeping indictment of food aid. As one who formerly had responsibility in the Federal Government for the planning and administration of food aid--the Food-for-Peace Program, I am aware that there is truth in many of these charges. But it is not the whole truth. This conference deserves a balanced assessment of food aid and Mr. Krauss has not provided it.

There are successes as well as disappointments in the history of food aid. I give you two examples, two out of a much larger number. One is the experience of the so-called Bihar Famine in India in 1966-67, the famine that didn't happen. It didn't happen because the United States provided, under Public Law 480 known as Food-for-Peace, 14 million tons of food grain. As a result of two consecutive years of drought, food-grain production in India had fallen 18 percent short. The drought was worst in Bihar Province, north of Calcutta. The American press reported that 15 million people were threatened with starvation.

As a condition of receiving food aid from the United States, India changed her agricultural price policy, which had been to hold prices low at the farm level, supposedly in the interest of food consumers, with the unintended consequence of reducing food production. Alarmed by the food crisis, India speeded up the Green Revolution, increased irrigation, and stepped up her agricultural extension system. By any measure this was a successful experience in the use of food aid. Instead of widespread starvation, the death rate fell. Instead of building

India into a situation of dependency, it was instrumental in expanding agricultural production. When the rains returned, the food aid was terminated. India's progress in food production since that time has been phenomenal. From before the Famine until 1982-83, wheat production trebled. This example should be placed alongside Mr. Krauss' gloomy assessment if a balanced appraisal is to be provided.

A second illustration is the feeding of Chinese refugees in Hong Kong after the revolution in Mainland China, and particularly during the famine that resulted from Mao Tse-Tung's disastrous Great Leap Forward in 1958. Refugees poured into Hong Kong by the thousands, many with nothing but the clothes on their backs. Emergency food aid was provided, under Food-for-Peace, by a cooperative effort involving the United States Government, the Lutheran World Service, and the Government of Hong Kong. Refugees were given food aid in the form of warm, fresh-cooked noodles, made from American wheat. Through the Lutheran World Service, they were also given health care and vocational training. Most of these people, who were able-bodied and industrious and who had come to a place where language and culture were familiar, quickly became self-reliant. Thereupon the program was terminated.

I could add other examples. I think our food aid to Ethiopia was successful in that it saved many lives despite very trying circumstances: a war underway, inadequate transportation, and an insufficient cadre of local officials. This may be a crisis which, as in India 30 years ago, serves to alert the country to the need of agricultural development, food security, and family planning.

What we need in assessing food aid is an examination of all the evidence, the good as well as the bad. We need the truth, not the half-truth.

8

Farm Structure, Productivity, and Poverty

Alain de Janvry

I. Introduction

The optimality of alternative farm sizes and tenure forms has been intensively debated on economic, welfare, and ideological grounds. Efforts at altering sizes and tenure farms through land reforms have at one time or another been legislated in virtually every country. And, indeed, ambitious land reform programs have, in several instances such as Taiwan and South Korea, established the preconditions for successful agricultural and industrial growth and greater equity in the distribution of income. Yet, few land reforms are occurring at the moment in spite of poor agricultural production performance, persistent extensive rural poverty, and the need to expand domestic markets for industry in a majority of Third World countries. There is substantial confusion and fatigue in the international development assistance community as to the worthiness of future land reform initiatives. Land reform seems to have lost its appeal as a tool for growth with equity (Adelman) and to be increasingly politically difficult to achieve (de Janvry). Lehmann has pronounced land reform officially dead and Ruttan recently concluded that "it is unlikely in the next several decades that political and economic developments will converge, as they did in a few areas in Asia and Latin America in the first two decades after World War II, that will provide assistance agencies with an opportunity to play a significant role in efforts to reform land tenure institutions or to achieve a more equitable distribution of land resources."

The thesis of this paper is that land reform remains a potentially extremely important instrument of economic development but that it is difficult to use because the conditions under which it can be successfully applied today are more narrowly delineated than they were in the past. The ideal condition for a redistributive land reform that creates family farms out of larger farms (with or without tenants) to achieve both productivity gains in agriculture and a reduction of rural poverty is when there exists a negative relation between total factor productivity and farm size and a positive relation between total household income and farm size. The objective of this paper is an attempt to identify the particular structural, economic, and social context under which these relations hold. Identifying the conditions under which land reform can meet with success should help make it a more effective instrument when and where it applies.

II. The Farm Size-Factor Productivity Relation

Since the success of a redistributive land reform depends on the existence of a strong inverse relationship between total factor productivity and the size of operational farm unit, it is important to first establish the theoretical conditions that explain the occurrence of this relation and the conditions that tend to weaken or reverse it.

1. Theoretical Determinants of an Inverse Relation

Starting with land productivity, there are three principal reasons why an inverse relation is expected to hold with farm size. One is a technological explanation according to which there are diseconomies of scale with farm size. Most studies, however, show that returns to scale in agriculture are approximately constant with respect to inputs actually used (Berry and Cline). Another explanation is that factor prices change with farm size—in particular labor, since small farms have access to cheaper (imputed) labor than farms which hire on the labor market. Finally, there is a behavioral explanation for the underuse of land on the larger farms in terms of semifeudal or non-profitmaximizing forms of landownership. Let us look at these last two explanations.

1.1. Labor Market Dualism and Supervision Costs

Labor costs tend to increase with farm size. This is due to three types of reasons. One is that family farms can make use of several labor categories for which the market fails: children, women in the reproductive periods of their life cycles, and elderly family members often have zero opportunity cost on the labor market. The marginal productivity of this captive labor will thus tend to be depressed way below the marginal productivity of labor on net-hiring farms, limited only by the trade-off between leisure and consumption but with a low marginal utility for leisure at low-income levels (Chayanov). A second explanation is that, even for household labor categories which have a positive opportunity cost on the labor market, the marginal productivity of labor tends to be below the wage paid on net-hiring farms due to avoidance of job search and transportation costs, risk of not finding employment off farm relative to the certainty of employment on the family farm (with, however, yield and price risks), and preference to work at home. Finally, family labor is more motivated to work harder than hired labor since it is a residual claimant on farm surplus. Family members thus work with maximum effort without supervision, while the work effort of hired laborers increases with supervision. If family members perform the role of supervising hired labor, output per acre will decline with operational farm size for a fixed family size and a given quantity of hired labor per acre since per-acre input of effective labor declines (Feder). The result is that the marginal productivity of labor on family farms is below that on net-hiring farms where it is equal to the market wage.

With no land market and no land rental market, theory thus tells us that the inverse relation between land productivity and farm size will be stronger if (1) labor market failures and labor captive to the household are more prevalent, (2) transaction costs on the labor market are higher, and (3) the labor effort of hired laborers is more significantly affected by supervision.

1.2. Objective Functions on Larger Farms

The other reason which creates an inverse relation between land productivity and farm size is the objective function that sometimes prevails among large traditional farm owners. When these farms are held by semifeudal landlords who do not manage their estates for the sake of profit making but hold land as a nonmarketable family asset or as a source of prestige, the marginal productivity of at least some factors tends to be substantially above opportunity cost. Land is also often used for speculative purposes or as a hedge against inflation in diversified portfolios where land is a low return-low risk asset to which little managerial attention is given, while urban and financial investments are the high return-high risk assets to which managerial efforts are allocated.

Aside from objective functions, land prices also tend to decline with the size of holdings. This is due to economies of scale in buying large tracts (Bhalla) and to a greater demand for small plots of land which can provide the large number of small farmers with opportunities to expand their farms.

We derive from this that the inverse relation between land productivity and farm size will be stronger when there is greater prevalence of semifeudal and nonprofit maximization behavior among large landlords and when land prices tend to decline sharply with size of ownership unit.

2. Factors that Weaken or Reverse the Inverse Relation

In addition to the eventual reversal of the factors mentioned above, there are several reasons why the inverse relation between land productivity and farm size may disappear or be reversed. They are as follows.

2.1. Lower Credit Costs for Larger Farms

In terms of expected returns from a loan to the lender, size of collateral and loan size are substitutes for the level of interest rate charged.

Consider, first, a loan to an owner-operator with ownership unit of size, OW. The expected return, E(L), from a loan, L, is

$$E(L) = (1 - P)\, L\, i + P(C - L) - K$$

where

P = probability that the loan will fail to be repaid
i = interest rate charged
$C = \bar{c}\, OW$, the collateral on the loan which is a given fraction $\bar{c}$ of OW

and

K = fixed transaction cost.

For a loan of given size $\bar{\ell}$ per acre of OW ($L = \bar{\ell}$ OW), the expected value of the loan for the lender is

$$E(\bar{\ell}) = (1 - P)\,\bar{\ell}\,i + P(\bar{c} - \bar{\ell}) - \frac{K}{OW}.$$

For a given expected return, e,

$$i = \frac{e - P\,\bar{c} + P\,\bar{\ell} + K/OW}{(1 - P)\,\bar{\ell}}.$$

The lender will then be willing to offer a lower interest loan:

1. In exchange for higher collateral per acre of ownership unit, $\bar{c}$.
2. To farmers with larger ownership units, OW.
3. On larger loans per acre of ownership unit, $\bar{\ell}$.

Larger farmers satisfy condition 2. If, in addition, they are willing to offer more collateral on hectares owned and are demanding larger loans per hectare, access to loan capital will be cheaper for larger farmers.

Next, consider the case of a loan made to a farmer who owns-operates OW and additionally rents OP - OW where OP is the size of the operational unit. Only OW can serve as collateral in the amount $C = \bar{c}$ OW. For a loan of a given size, $\bar{\ell} = L/OP$ per acre of operational unit, the expected value of the loan for the lender is

$$E(\bar{\ell}) = (1 - P)\,\bar{\ell}\,i + P(\bar{c}\,\frac{OW}{OP} - \bar{\ell}) - \frac{K}{OP}.$$

For a given expected return, e,

$$i = \frac{e - P\,\bar{c}\,OW/OP + P\,\bar{\ell} + K/OP}{(1 - P)\,\bar{\ell}}.$$

The lender will then be willing to offer a lower interest

loan to that farmer if:

1. He owns a larger share of the operational unit (OW/OP).
2. The collateral per acre of ownership unit is raised (c).
3. The size of the loan per acre of operational unit is greater ($\bar{\ell}$).
4. The size of operational unit is greater for given OW/OP.

If smaller farmers tend to rent a larger share of their operational unit,[1] condition 1 implies that interest rates decline with farm size.

There are thus several good reasons why, even with equilibrium conditions for lenders, the cost of capital can be expected to be lower for larger farmers. This is confirmed, for instance, in India where the rate of interest charged decreases from 17.3 percent on farms of less than 5 hectares to 11.8 percent on farms larger than 25 hectares (Bhalla). The result is to induce a greater degree of capital deepening on the larger farms.

Whether capital deepening on the larger farms cancels, or not, the inverse relations between land productivity and farm size depends on the bias of technological change. If investment occurs in laborsaving technological change, such as mechanization and particularly tractors, the inverse relation is weakened. Weakening of the inverse relation results from the fact that, even if laborsaving technology is yield neutral, by being laborsaving it also saves on time allocated to labor supervision and, hence, increases the effectiveness of hired labor for given farm and family size. If investment occurs by contrast in landsaving technology, such as fertilizers, new seeds, and irrigation (the so-called Green Revolution), land productivity likely will increase with farm size. In this case the stronger the inverse relation between interest rate and farm size, the stronger the positive relation between land productivity and farm size. This positive relation is, however, weakened by rising supervision costs if landsaving technological change is employment creating for given farm and family size.

2.2. Land-Rental Market with Interlinked Factor Markets

If there exists a perfect land-rental market, the distinction between ownership and operational holding becomes important. We have seen that, as the size of operational holding increases, land productivity tends to decline as labor costs increase and land productivity tends to increase as credit costs decline. If, however, landlords rent their estates in small plots of land and give credit to their tenants, they are able to accumulate the advantages of a large ownership unit to gain access to cheap credit and of small operational units to capture cheap tenant family labor as well as other non-marketable inputs which the households control (Binswanger and Rosenzweig). Since landlords usually have better information on tenants than do moneylenders, they can avoid problems of moral hazards and of adverse selection in screening when they lend to their tenants and thus can do so at low interest rates. The size of the plots which landlords give to tenants depends on family size, the ownership of work animals and machines, and the managerial skills of tenants. Since there is an inverse relation between the tenant's effort per acre and plot size, the landlord can rent out plots of a size no greater than what is needed to insure to the tenant a level of utility equal to what he could have obtained in an alternative occupation (Braverman and Srinivasan).

Thus, even with impediments to land sales and to the transaction of family labor, work animals and machines, and managerial and supervisory skills, tenancy and the interlinking of credit and land rental allow the equalizing of factor ratios across sizes of ownership holdings. With tenants benefiting from the credit advantages of landlords, there may no longer exist a relationship, either negative or positive, between size of ownership holding and land productivity. The less landlords are able or willing to provide credit to their tenants, the more land productivity would decrease with size of operational holdings (as both credit and labor costs increase).

Linking of credit and land-rental markets thus increases output and landlord income. If landlords can always hold tenants' incomes down to their opportunity costs, effective linking will lead them to reduce plot size and increase the number of tenants (Cheung). Thus, only a redistributive land reform can possibly increase the tenants' welfare by destroying the landlord-enforced inverse relation between peasant income and plot size.

2.3. Continuous Rapid Landsaving Technological Change

Numerous empirical studies of the diffusion patterns

of Green Revolution technology have shown that larger farmers have been adopting modern varieties sooner and more extensively than smaller farmers (Lipton and Longhurst). This has been attributed to fixed setup costs (learning, locating markets, and training labor) which discourage adoption by small farmers and to their behavior toward risk. If, as expected, absolute risk aversion decreases with farm size while relative risk aversion increases, new technologies which are risk reducing (e.g., insecticides) will be used more on larger farms while those which are risk increasing (fertilizers) will be used less (Just and Zilberman). With empirical evidence indicating that high-yielding varieties may well be risk reducing (Lipton and Longhurst), behavior toward risk would then explain greater adoption on larger farms. The diffusion pattern of land-saving technology thus tends to create a positive relation between land productivity and farm size.

Later studies of the Green Revolution have, however, demonstrated that small farmers are able to subsequently catch up in adopting if they have not been displaced in the process by falling product prices. The lag in adoption depends on the institutional framework which determines relative access to credit, information, and modern inputs as well as on technological biases in research. It is also explained by differential credit costs and behavior toward risk across farm sizes. In India, 10 years after initiation of the Green Revolution, the relation between adoption of modern varieties and farm size had essentially disappeared for wheat and, in most states, for rice as well (Baker and Herdt).

If there is a continuous rapid flow of new landsaving technologies, the lag in adoption relative to the best technology tends to be reproduced, and the inverse relation between land productivity and farm size can be permanently negated.

2.4. Institutional Rents

Most Third World governments tend to intervene heavily in the determination of agricultural prices, principally through taxation of export crops in order to generate government revenues (e.g., parastatals in Africa) and through overvaluation of the exchange rate and price controls to cheapen staple foods. Product price distortions maybe partially or totally compensated through input subsidies, but the distribution of subsidies tends to occur through the forces of the political economy. Thus, the market is used to tax agriculture and the political economy to selectively compensate particular producers and activities. This process of give and take has its logic in terms of clientelistic gains for the state and has been analyzed as the theory of institutional rents by de Janvry and Bates.

Because of the correlation between landownership and political power, larger farmers are usually able to monopolize access to institutional rents which come to them under the form of rationed subsidized credit, tax write-offs on capital investments (such as tractors), technological biases toward their specific production systems and factor price ratios, and favorable location of public work projects such as irrigation and roads. The distribution pattern of institutional rents across farm sizes, particularly if they support the adoption of landsaving technological innovations, thus tends to create a positive relation between land productivity and farm size. The more distorted the economy and the more unequal the distribution of political power in the allocation of compensatory institutional rents, the more the inverse relation tends to be negated. Programs of integrated rural development that attempt to improve small farmers' access to institutional rents can thus be seen as efforts to counteract this tendency.

3. Robustness of the Relation and Other Mediating Factors

Many empirical studies have indeed confirmed the existence of an inverse relation between land productivity and farm size. Berry and Cline, after correcting for declining land quality with farm size, find that the land productivity on smaller farms is at least twice that observed on larger farms. In Colombia they observe a ratio of 2 to 1; in northeastern Brazil, of more than 2 to 1; in India and Malaysia, of over 2 to 1 (Berry). Most of these data are, however, from the 1960s and early 1970s before productivities were significantly affected by antifeudal land reforms in Latin America and by the Green Revolution in Asia. There is mounting evidence that this inverse relation has been weakened and sometimes reversed, at least on the medium and large farms. This has been observed in the Punjab by Roy and by Bhalla, in West Bengal by Barbier, and in Mexico by Burke (see, also, Feder, Just, and Zilberman). Thus, empirical evidence does show that both positive and negative relations can be observed.

We bring new empirical evidence on the relationship between land productivity and farm size by analyzing data for 15 countries collected by the Food and Agriculture Organization of the United Nations around 1970 (see Cornia). We postulate the hypothesis that there exists a discontinuity in this relationship between farms below the size of a family farm (AF) and farms above that size with few cases of a significant negative relationship for farms larger than family size. A strong negative relation in subfamily farms is explained by a rising share of tradable labor in total family labor and by increasing marginal utility of leisure as farm size increases. Disappearance of this relation for larger than family farms would indicate that biased access to capital, landsaving technology, and institutional rents overwhelms rising costs of effective labor and rising marginal utility of leisure. We correct for differences in land quality by using cultivated area instead of farm size whenever the former is less than the latter (i.e., the land utilization index is less than one). The estimated equations are presented in Appendix Table 1.

As Table 1 (Tables at the end of chapter) shows, in 11 of the 15 countries analyzed, there is no significant relation between land productivity and farm size for farms above the size of a family farm. Of these 11, 6 have a significant negative relation for farms smaller than a family farm. Only four countries have negative relations for farms larger than a family farm. Among these, there is substantial evidence that the negative relation has disappeared in India as a sequel to the Green Revolution. The other three are all land-abundant African countries where landsaving technological change has generally not been introduced. From this analysis, we thus conclude that:

1. There generally exists a discontinuity in the relation between land productivity and farm size around the size of a family farm.
2. Few countries evidence a negative relation between land productivity and farm size for farms larger than a family farm, particularly in land-scarce economies where landsaving technological change has been important.
3. In 5 of the 15 countries, the R^2 in the relation between land productivity and farm size is low (below .5), indicating that a number of other

variables contribute to the determination of land productivity.

Since redistributive land reforms would likely avoid creating farms smaller than family size, the implication is that there are relatively few situations where strong productivity gains will result from land reform in the existing institutional context of access to capital, technology, and institutional rents. It is likely that this conclusion would be even stronger if the analysis were based on more recent post-Green Revolution data, in India in particular.

4. Types of Land Reforms in Relation to Potential Productivity Gains

Figure 1 (Figures at the end of chapter) shows five archetype relations between land productivity and farm size derived from the above theoretical discussion and empirical analysis (see the Appendix for a formal model from which these relations are derived). Defining land reform as a transition between two states of the land tenure system and defining a state of the land tenure system by the combination of farm size (e.g., large net hiring, medium net hiring, and family farms) and of land productivity-farm size relations (semi-feudal, profit maximizing without landsaving technology, ownership with tenancy and interlinked land rental-credit markets, and profit maximizing with landsaving technology), we obtain 12 types of land reforms. Historical illustrations of the main types of reforms are given in Table 2 (Tables at the end of chapter).

Antifeudal land reforms (1-4 and 1-5) are the main types of reforms that occurred in Latin America. They were reforms without significant redistribution where the shifts (1-4) and (1-5) were obtained through threats of expropriation to induce private evasive actions. In reforms (1-4), the threats of expropriation came from the imposition of minimum productivity standards to induce profit-maximization behavior creating so-called "Junker" farm operators out of former semifeudal landlords. In reforms (1-5), maximum sizes for ownership units were established (80 irrigated hectares in Chile; 200 in Mexico) while providing landlords with enough lead time to engage in private sales or redistributions within the family to avoid expropriation. Antifeudal reforms were the most important and the most successful reforms in inducing productivity gains even though redistributive expropriations were minimal. When they occurred in conjuction with the Green Revolution and effective programs of agricultural development, as in Colombia (1-6) and Mexico (1-7), they were able to substantially increase agricultural output. Today, with the virtual disappearance of feudalism, these reforms have been successfully completed.

Redistributive reforms that create family farms through actual expropriations end in state 2 of the land tenure system. They have also generally originated from feudal systems (Taiwan and South Korea) and thus achieved susbstantial productivity gains (1-2). Note that productivity gains starting from profit-maximizing farms are smaller and can be negative in a post-Green Revolution situation (6-2). This type of land reform, like that promoted by President Echeverria in modern northwestern Mexican agriculture, is the most likely to result in declining land productivity unless accompanied by credit reforms and effective technical assistance.

Antitenancy reforms (8-4, 9-5, 10-6, and 11-7), which ban free tenancy arrangements without reforming the ownership structure, are the least likely to succeed in raising agricultural output. This is because a land rental market is desirable, whether land is owned unequally or not, to adjust the size of operational unit to ownership of nontraded or imperfectly traded factors of production such as family labor, work animals, labor supervision, and entrepreneurship. Since tenancy arrangements are institutional substitutes for markets that fail or are distorted, eliminating tenancy without perfecting market alternatives will create substantial inefficiencies. Since private evasive actions to antitenancy reforms result in the eviction of tenants, efficiency losses are also accompanied by rising rural poverty. As the disastrous Argentinian antitenancy reforms of the 1940s show, nonredistributive antitenancy measures should never be an element of land reform.

Redistributive antitenancy reforms (8-2 and 10-2), whereby land is distributed to the former tenants, are equally problematic from an efficiency standpoint. If the land and credit markets are interlinked and if landsaving technological changes are important, tenants will only benefit from redistribution if the credit market is reformed to provide them with similar terms to the ones they could obtain through the landlords. Private initiatives by landlords to avoid the reform have almost inevitably led to massive evictions of tenants; and these reforms, of course, failed to benefit the landless. If properly accompanied by credit reforms and managed to control massive evictions, redistributive tenancy reforms are, however, the key to reduce rural poverty among tenants since they break the vicious circle of rising productivity (e.g., through adoption of landsaving technology) leading to falling plot size, thus allowing landlords to maintain tenants at their reservation wage in a principal-agent relationship while capturing all the gains from technological change.

5. Implications for Land Reform

Before drawing implications for land reform, we must recall that the aggregate output effect of land reform depends on the relationship between the land tenure system (farm size; social relations and technology) and total factor productivity at social prices, not land productivity (Berry and Cline). The negative relation between total factor productivity at social prices and farm size tends to be weaker than that between land productivity at market prices and farm size if (1) labor productivity increases sharply with farm size because of laborsaving capital investments; (2) labor markets are closer to full employment so that the social opportunity cost of family labor is not zero; and (3) markets for family labor, animal, machinery, labor supervision, and entrepreneurship are more perfect so that there are less captive resources in rural households and lesser transaction costs.

With at least the latter two conditions unlikely to hold in the Third World, we can derive the following implications for the productivity effects of land reform.

1. Antifeudal land reforms were the easy phase of land reform since they could easily achieve efficiency gains and were politically easy to implement since they could achieve their productivity objectives eventually without significant expropriations and redistributions. They are a largely exhausted possibility today since feudal relations have practically disappeared as a consequence of land reforms and the development of capitalism in agriculture.
2. Nonredistributive antitenancy reforms are bound to fail unless substantial market reforms substitute for the logic of existing institutional arrangements. Redistributive tenancy reforms thus need to be complemented by other reforms,

particularly on the credit market. While they are difficult to implement, they are key to raising the welfare of tenants above their current reservation wages.
3. Redistributive land reforms need to be part of comprehensive rural development programs (an observation made by Warriner in the 1960s), particularly if landsaving technological innovations have started to diffuse extensively.
4. With the rising importance of capital investments in landsaving technological change, alternatives to redistributive land reforms increasingly exist to raise agricultural productivity. Thus, effective agricultural development programs can provide an alternative to land reform in terms of productivity gains. Conversely, this means that achieving productivity gains in a postfeudal, landsaving, capital-intensive agriculture via redistributive land reforms will be increasingly difficult since they require to transform the institutional context that determines access to capital and technology.

III. The Farm Size-Rural Poverty Relationship

Land reforms have income objectives in addition to productivity objectives. Income goals can be set by dominant groups for the sake of legitimation or domestic market expansion, or in response to pressures from landless rural workers and small farmers. Here, again, the ideal situation for a redistributive land reform to reduce rural poverty is when there exists an inverse relation between incidence of poverty and farm size. We establish here the theoretical conditions that establish this inverse relation and the factors that tend to weaken or reverse it.

1. Theoretical Determinants of an Inverse Relation

We can easily construct a simple theoretical model where the amount of land owned both differentiates social classes in agriculture and defines the level of welfare according to an inverse relation between poverty and farm size (see the Appendix).

Following Bardhan, we can define five social groups based on the direct use of family labor in home production (L), family labor hired out (H), and wage labor hired in

(N). Family labor is composed of both captive labor, L_1, and tradable labor, L_2. Tradable family labor is needed to supervise hired labor so that effective labor increases with supervision (S). The five house hold types are as follows:

Household type	Family labor				Hired labor
	Captive	Tradable			
	L_1	L_2	S	H	N
Landless worker	0	0	0	+	0
Subfamily farm	+	+	0	+	0
Family farm	+	+	0	0	0
Rich farmer	+	+	+	0	+
Capitalist farmer	0	0	+	0	+

If each household maximizes a utility function in income and leisure, we can see that the ranking of households from subfamily farmer to capitalist farmer is determined by the initial amount of land owned and that the maximum level of utility reached by each household increases with the size of the ownership unit. To establish this, it is enough to accept the following three rules:

1. Family labor is cheaper to use than hired labor since it does not require costly supervision. This transaction cost in access to hired labor implies that it is always better to fully use family labor before starting to hire in.
2. The marginal productivity of family labor in the home plot is greater than the market wage over at least some minimal range. This implies that some family labor will be assigned to the home plot before being hired out provided differential gains between home and wage labor cover start-up costs.
3. Capital can be borrowed with land as collateral, i.e., in an amount proportional to the size of ownership unit. Capital can, in turn, be used to rent additional land, hire labor, and cover fixed capital costs.

As the size of ownership unit increases from the smaller farm size, these three rules establish the social class ordering from subfamily farmer to capitalist farmer.

It is also obvious that welfare increases with farm size since any social class could always opt for the labor strategies of the classes below itself but does not in maximizing utility. This simple theoretical model thus establishes an inverse relation between the incidence of poverty and size of ownership unit.

The level of wage on the rural labor market is affected by the land tenure system in the sense that it depends upon the amount of land controlled by semiproletarian (subfamily farm) households and the relative contributions to total labor supply of full versus semiproletarians. If peasants lose access to land, two opposite effects can change the level of wages. An increasing labor supply will depress wages with the results that net-hiring farmers will gain. Net employers thus benefit from expropriating peasants and consolidating the land. If, on the other hand, a reservation utility level exists for peasants (determined, for instance, by expected returns from migration) falling access to land will raise the level of wages. Net employers thus should gain from protecting continued access to land by peasants, for instance, through modest redistributive land reforms that create subfamily farms or through rural development programs to enhance the productivity of labor in peasant farming.

2. Factors that Weaken or Reverse This Inverse Relation

While the inverse relation between poverty and farm size is established on the basis of simple behavioral principles, there are several factors that weaken or reverse it. A redistributive land reform will be less effective in reducing poverty when these factors prevail.

2.1. Ownership of Other Assets

Landless households may own assets other than land such as draft and other animals, education (human capital), durable producer and consumer goods, financial assets, and jewelry. Some of these assets can serve as collateral and, hence, for access to rented land. Yet, except for financial assets, the value of the other assets as a source of collateral is less than land because they usually lack one or more of the fundamental requirements of an ideal collateral identified by Binswanger and Rosenzweig: Appropriability by the lender, absence of collateral-specific risk (moral hazard and lack of insurance), and accrual of the returns to the borrower during the loan period. Even without collateral value, these other assets are income generating and thus weaken the relationship between ownership units and poverty.

2.2. Land Rental and Interlinked Markets

It is precisely the ownership of nonland productive assets by landless households and imperfect rental markets for these assets that leads to tenancy arrangements under the form of cash rental or sharecropping. They include many categories of family labor, draft animals and machines on a long-term basis, and managerial and supervisory skills. With tenancy arrangements, access to land is more equal than landownership. As we have seen, if the landlord has the ability of reducing the size of operational units to the tenant's reservation wage, tenancy arrangements will not increase household welfare. If, on the other hand, product shares can be bargained because potential tenants own nontradable factors, the tenant's income can be above his opportunity cost as a landless worker or a small owner-operator farm.

2.3. Land Not the Scarce Factor

There are situations where the limiting factor on production is not land but labor or productive capital. This is still the case in many African countries where access to land is determined by customary rights and where the amount of land cultivated by a household is determined by its capacity of mobilizing labor and laborsaving technology. This is typical of slash-and-burn agriculture. While these situations are increasingly rare and suffer from the tragedy of the commons, they are conditions where the relation between individual access to land and poverty does not hold in the short run even though it does in the long run through the externalities that increased land use imply on the community.

2.4. Importance of Nonfarm Labor Incomes

If expected wages from off-farm employment increase, family labor is reallocated from the home plot to hiring out. At the limit, if wages are always greater than the marginal productivity of labor on the home plot, all noncaptive labor categories are reallocated to the labor market. Wages are thus a substitute for land in peasant household welfare, and rising off-farm income opportunities weaken the inverse relationships between farm size and rural poverty. They also reduce the minimum farm size which is necessary to reach the household's reservation income.

2.5. Mechanisms of Wage Determination

The mechanism of wage determination in agriculture establishes whether or not labor productivity gains on net-hiring farms and increases in the prices of the products sold by these farms are passed along to their workers under the form of rising real wage bills. If nominal wages are fixed in the short run, the real wage bill gains from rising productivity will be greater if the employment effect and the overall deflationary effect, due to increased product supply with an inelastic demand, are strong. If there is full employment, the real wage bill will only increase with productivity gains if there is a strong employment effect and if the overall deflationary effect is greater than the price decline in the particular crop considered. Since these conditions are generally unlikely to hold, the welfare of landless workers will tend to decline with that of their large farmer employers when technological change depresses farm prices due to an inelastic demand for food (de Janvry and Sadoulet).

2.6. Size of Household and Life-Cycle Stage

As observed by Chayanov, household poverty measured by the degree of "self-exploitation" (the intensity of work) increases in the early stages of the household life cycle as the number of dependents per worker increases. It later decreases as children enter into the labor process. The earlier children can be used productively, for instance in substituting for adult labor in petty tasks, and the longer they can be retained as contributors to the household, the better off a household of a given family size will be. For as long as children can be used productively, the larger the family, the weaker the relation between poverty and farm size.

3. Empirical Information

Appendix Table 2 provides information on household income and source of income by farm size from 10 farm surveys in seven countries. Regression results in Table 3 (Tables at the end of chapter) show that there is in all cases a strong positive relation between farm income and farm size. Access to land is thus the best guarantee against rural poverty; and redistributive land reform is, consequently, the most effective antipoverty instrument.

The results in Table 3 also show that there generally exists a stronger relation between farm income and farm size than between total income and farm size, indicating that access to nonfarm sources of income (off-farm employment opportunities, in particular) is an important substitute for land in raising household income. In situations where landless and marginal farmers (smallest farm categories) have good access to nonfarm income opportunities, their income can be above that of small farmers (Peru, Bolivia, Mexico, India, and the United States). The Mexican data in Appendix Table 2 show, for instance, that in the 0-4 hectares farm interval, the more proletarianized households (F_2) derive a household income which is 5 percent higher than that of the less proletarianized households (F_4 and F_1 combined).

Fitting a quadratic function to the relation between nonfarm income and farm size indicates that there is in most situations a strong negative nonlinear relation between the two. This is what explains the observed nonlinear relation between total income and farm size mentioned above. Figure 2 (Figures at the end of chapter) represents the observed relations. The fact that the level of total household income observed on farms of size FS* is lower than that on farms smaller than FS* and that of the landless is an interesting puzzle. Why would these households not reduce the size of their farms or become landless to capture higher income levels? There are two possible lines of explanation of this puzzle.

One is that the observed income levels are equilibrium ones and that there is no logic to migrate or reduce farm size. This can be due to several reasons such as the following:

1. Transition costs from migration are equal to the difference between Y and Y*.
2. The cost of living is lower on small farms (FS*) than in urban or suburban environments so that the purchasing power of Y* is not inferior to that of Y.
3. The level of risk associated with Y is higher than that associated with Y* since small farms offer a more diversified source of income. After

discounting for risk, the two levels of income are perceived as equally desirable.

4. The difference between desired and actual family size is greatest on farms of size FS* due to malnutrition and poverty.

The other line of explanation is that the observed income schedule reflects a disequilibrium situation but that small farmers (FS*) do not have the opportunity of access to sufficient off-farm employment to migrate or reduce farm size. Migration is thus blocked by lack of access to off-farm employment, and causality runs from off-farm employment to farm size. With probabilistic access to off-farm employment at an institutional level of wage, no income sharing among landless (as opposed to Todaro's model), and social networks that transmit the information to potential migrants on small farms when a job is available for them, the waiting line of unemployed is located on the small farms which are thus the reservoir of surplus labor. These small farms are abandoned when good off-farm income opportunities arise. Income on the small farms can thus be increased by greater access to either land or employment opportunities.

Table 4 (Tables at the end of chapter) shows results obtained from a computable general equilibrium model for India that evidence the key roles of employment opportunities, wage levels, and food prices in mediating the relationship between rural incomes and farm size (de Janvry and Subbarao; de Janvry and Sadoulet). Total household income in the base period (1979) increases with farm size but is lower for small farmers than for the landless. Simulation of a 10 percent increase in agricultural output with flexible prices shows that the landless are the ones who gain most from the resulting increased employment opportunities and falling food prices. Small farmers also gain since they hire labor out and are net buyers of food. Large farmers only benefit from rising output if prices are fixed by either export demand or government price guarantees and stockpiling. This explains the demands of the large farmers' lobbies and their current success in obtaining implementation of such policy interventions. Landless workers and small farmers benefit more from constant nominal wages, a likely phenomenon in the short run only, than from wages that respond to supply and demand on the labor market. In this latter case, falling product prices depress the value of marginal productivity of their labor and their nominal wages. The opposite occurs when prices are fixed since nominal wages then increase with productivity gains. Fixed real wages, a plausible long-run condition in Indian agriculture due to the secular permanence of surplus labor (Lipton), rob the landless and small farmers from benefiting significantly from the deflationary effects of output growth.

Finally, it is important to note that, under conditions of surplus labor and land scarcity, access to even a small plot of land makes a significant contribution to the reduction of poverty. This is particularly true if this land is irrigated and the income-generating capacity of this plot of land is not excessively risky. When both landless and landed workers participate in the same labor market and if the labor demands on the home plot do not compete with peak labor demand on the labor market (when wages are the highest), functional dualism through access to a small plot of land is an effective instrument of poverty alleviation. Land reforms that distribute subfamily plots of land can thus reduce rural poverty for as long as a fringe of landless workers remains to protect wages from falling below the reservation wage of landless workers.

4. Types of Land Reform and Poverty Alleviation

Figure 3 (Figures at the end of chapter) gives the relations between farm size and farm income, total household income with surplus labor, total household income with full employment, and tenant income that derive from both the above theoretical and empirical analyses. Defining social relations in agriculture as the intersection between a farm size (landless, subfamily farms, and family farms) and an income line (tenant income and total household income with and without surplus labor), we can identify how different types of land reforms will affect the class structure and poverty levels in agriculture.

Taking 1 as the reservation wage for tenants, we see how redistributive antitenancy reforms can benefit former tenants if access to land is sufficient (5-4) but not necessarily if it creates subfamily farms under conditions of surplus labor (5-2). Land reforms that transform subfamily farmers who had access to permanent employment in haciendas into subfamily farmers facing surplus labor on labor markets (3-2) will likely cause their total household income to decline. This was the case for the permanent workers in haciendas in Ecuador and Bolivia who were granted the plots of lands on which they lived but lost their permanent labor contracts with the landlord (Table 5) (Tables at the end of chapter). It is for this reason that land reform was generally opposed by the permanent workers in the haciendas of Ecuador. When the plots of land granted to reform beneficiaries are sufficiently large (3-4), as in Chile, total household income will, however, increase. Finally, the ideal antipoverty land reform is when landless agriculture workers are provided access to family farms (1-4). This occurred in Taiwan and South Korea but virtually never in the Latin American land reforms which benefited instead the landed permanent workers on haciendas (3-4).

We thus conclude that access to land via redistributive land reforms that create family farms remains the most effective antipoverty strategy, particularly if there is surplus labor. If land is scarce, access to even a small plot of land will reduce poverty if employment conditions are sufficiently favorable. In this case, attention must be given to creating off-farm income opportunities (2-3). It is important to realize, however, that land reforms will easily hurt several categories of rural poor, particularly the excluded landless who will lose employment opportunities and the former permanent workers of large farms if there is surplus labor and if they do not gain access to more land than they had before. Evasive actions by landlords can also lead to massive expropriation of sharecroppers, tenants, and permanent workers. The main welfare result of the Latin American land reforms has thus been to replace permanent by temporary workers on the net hiring farms, to sharply increase the number of subfamily farms and thus reinforce functional dualism, and to locate in rural towns an agricultural labor force available for temporary hire generally through the mediation of labor contractors. The production successes of antifeudal land reforms has been matched by their failure to alleviate rural poverty.

IV. Conclusion

The economic superiority of the family farm rests on labor market dualism and on existence of nontraded or imperfectly traded factors of production controlled by farm households. The total factor productivity superiority of the family farm will, consequently, tend to disappear when (1) surplus labor is eliminated, (2) labor market failures are erased which thus frees captive resources (e.g., female labor), (3) captive labor is no longer a productive resource (e.g., children who go to school instead of working on the farm), and (4) other factors controlled by the farm household can be traded, in particular managerial and labor supervision functions through the hiring of foremen or labor contractors. The total factor productivity effect of redistributive land reforms will thus be potentially more important when these conditions do not hold that is in poor, labor surplus, agrarian economies ridden with substantial market failures. Antitenancy land reforms are, however, likely to fail to achieve productivity gains in this context precisely because contractual arrangements tend to emerge in compensation for market failures.

When a rapid flow of landsaving technological innovations occurs, access to technology and credit through rural development programs is essential to protect the productivity superiority of the family farm. This requires specific research efforts directed at its farming systems and removal of disadvantages in access to credit resulting from transaction costs, tenancy contracts with imperfect land and credit markets, and exclusion from access to institutional rents.

Preservation of farms smaller than family farms depends crucially upon access to nonfarm sources of income and, in particular, on the level of wage relative to the reservation wage of landless workers. Under surplus and captive labor conditions, even access to a small plot of good quality land can significantly alleviate rural poverty.

While there are many situations where a redistributive land reform is unlikely to increase total factor productivity in agriculture, access to land remains the key determinant of rural income. Even when land reform thus loses its productivity rationale because the inverse relation between productivity and farm size has disappeared, it remains key for social welfare in the rural sector in the absence of intersectoral income transfers toward the rural poor.

TABLE 1

Relation Between Land Productivity and Farm Size in 15 Countries

$A \geq A_F$ / $A \leq A_F$	No significant relation	Significant negative relation	Number of countries
No significant relation	Barbados Mexico Peru Bangladesh Thailand	India	6
Significant negative relation	Sudan Syria Tanzania Burma Nepal Korea	Ethiopia Nigeria Uganda	9
Number of countries	11	4	15

Source: Appendix 1.

TABLE 2

Types of Land Tenure Systems

Pre-land reform \ Post-land reform	4, 6 junker	5, 7 farmer	2 family farms
1 Semifeudal	Bolivia, 1952- Ecuador, 1964- Colombia, 1968- Peru, 1964-1969	Mexico, 1934-1940 Chile, 1967-1973 Egypt, 1952-1966 India, 1950-	South Korea, 1950- Taiwan, 1951-1963
4, 6 junker		Peru, 1969-1975 Philippines, 1972-1979	
5, 7 farmer		Mexico, 1940-	Dominican Republic, 1962-
8, 9 tenants	Argentina, 1945-1955	India	Philippines, 1972-1979

Junker = large net-hiring capitalist farm.

Farmer = medium net-hiring capitalist farm.

TABLE 3

Relation Between Total Income, Farm Income or Nonfarm Income, and Farm Size (FS)[a]

Endogenous variable:	Farm income		Total income		Nonfarm income	
Exogenous variable:	FS		FS		FS	FS^2
Cajamarca (Peru)	18.5*	>	17.3*	>	-7.0	0.1
Garcia Rovira (Colombia)	47.8*	=	47.2*	>	0.6	b
South Bolivia	11.9*	>	5.2*	>	-16.0	0.5
Puebla (Mexico)	96.3*		100.1*	>	-13.3	1.2
Region IV (Chile)	461.0*		503.6*	>	-33.0*	5.9*
Sierra (Ecuador)	60.0*		85.3*	>	-5.2*	0.2*
Punjab and Haryana (India)	25.6*	>	22.6*	>	-95.3	7.4
Gujarat and Maharastra (India)	21.6*	>	18.1*	>	-20.8*	0.4*
Kerala and Tamil Nadu (India)	78.0*	>	73.0*	>	-29.0*	0.8*
Bengal and Bihar (India)	29.5*	>	19.1*	>	-21.5*	0.4*
United States	418.4*	>	393.8*	>	-197.8*	1.4*

[a]Regression coefficients between the three concepts of income and farm size.
[b]Insufficient number of degrees of freedom.
*Significant at the 95 percent confidence level.
Source: Appendix Table 2.

TABLE 4

Rural Income Structure in India

	Percent of HHS	Farm income	Agricultural wage income	Other income	Total income	10 percent output increase with flexible prices: Fixed nominal wages	10 percent output increase with flexible prices: Full employment	10 percent output increase with flexible prices: Fixed real wages	10 percent output increase with fixed prices: Fixed nominal wages	10 percent output increase with fixed prices: Full employment	10 percent output increase with fixed prices: Fixed real wages
acres	percent	rupees				percentage change in real total income					
landless	12.3	38	354	229	622	39.5	25.8	18.1	7.4	15.5	7.4
0.1-3.8	34.6	246	101	155	502	15.9	11.1	8.5	8.1	10.9	8.1
3.8-6.9	37.0	669	0	161	829	-8.0	-7.8	-7.5	8.9	8.2	8.9
6.9+	16.1	1,807	0	216	2,024	-27.3	-24.4	-22.5	9.4	6.6	9.4
						percentage change in wages and prices					
Nominal wages						0	-18.5	-25.1	0	12.6	0
Consumer price index						-23.7	-24.6	-25.1	0	0	0
Real wages						23.7	6.1	0	0	12.6	0

Sources:

A. de Janvry and K. Subbarao.
A. de Janvry and E. Sadoulet.

TABLE 5

Effects of Alternative Land Reforms on Household Incomes

Post-land reform / Pre-land reform	SF with surplus labor 2	SF with full employment 3	Family farm 4
Landless 1	-	+	++ Taiwan South Korea
SF with surplus labor 2		+ Employment creation	++
SF with full employment 3	- Ecuador Bolivia		+ Chile Egypt
Tenant 5	-	+	++ Philippines

SF = Subfamily farm.

The signs indicate the direction of change in household income.

Figure 1. Relation between land productivity and farm size

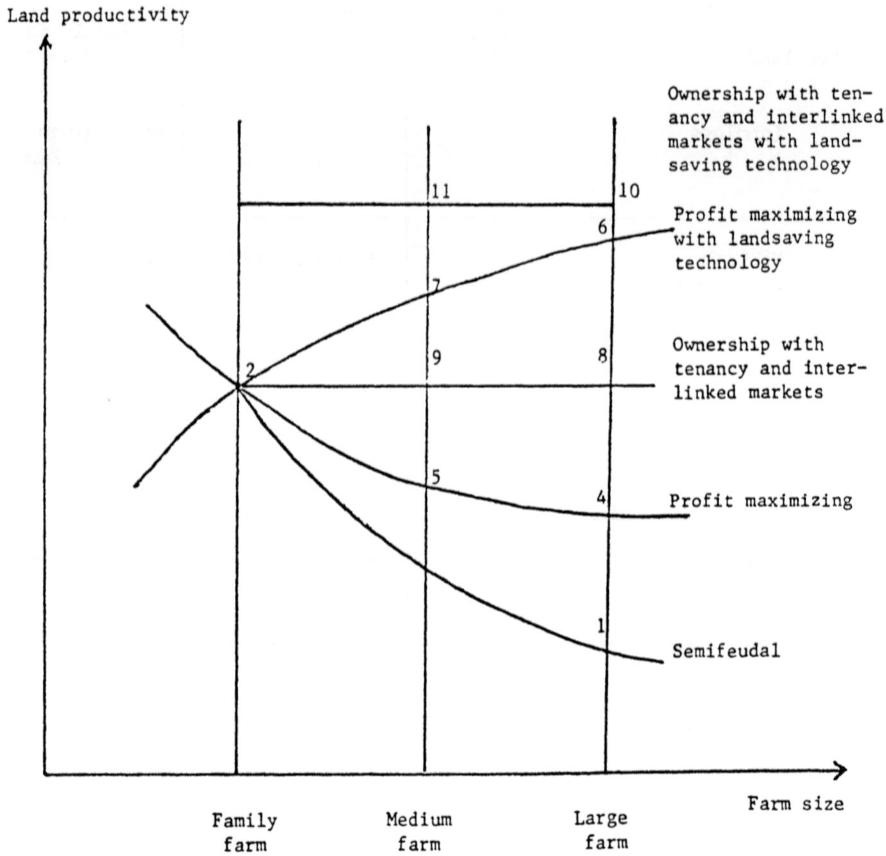

Figure 2. Relation between household income and farm size

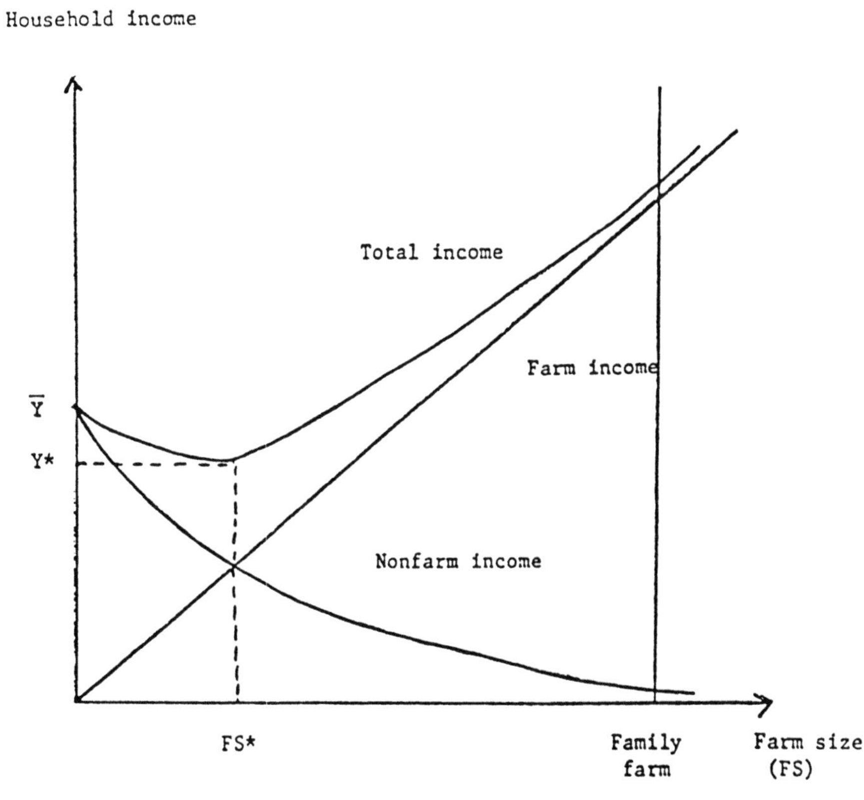

Figure 3 - Relation Between Household Income and Farm Size

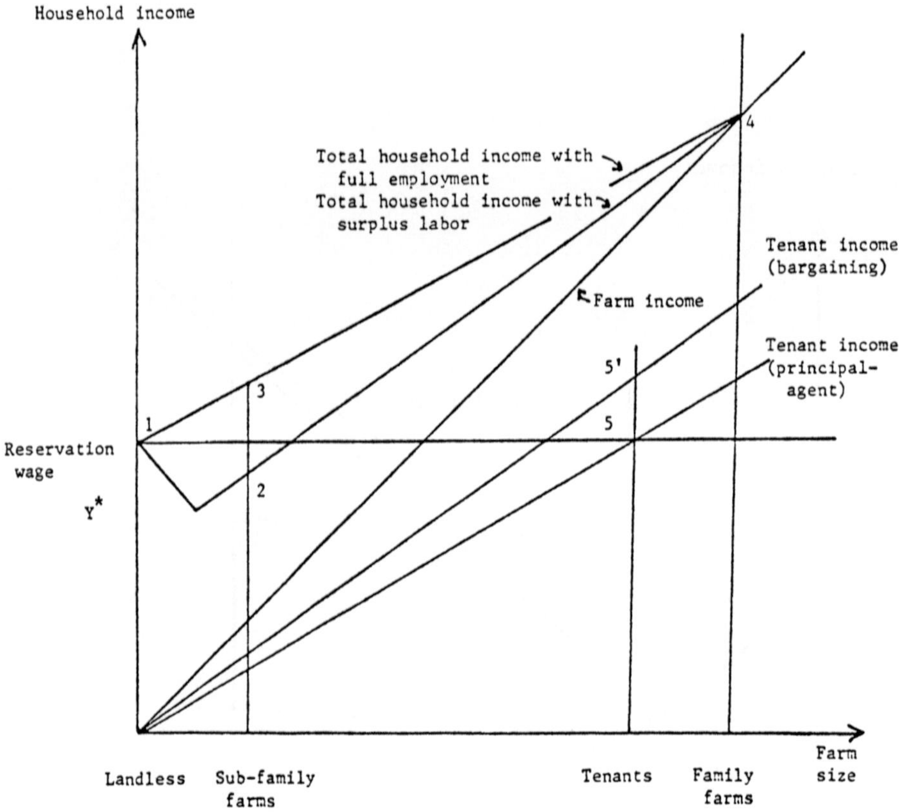

FOOTNOTES

[1]In the United States, large operational holdings have a higher share of rented land than do small ones. In Egypt, by contrast, smaller farms have a higher share of rented land.

BIBLIOGRAPHY

Adelman, I. "Growth, Income Distribution, and Equity-Oriented Development Strategies," World Development, Vol. 3, Nos. 2 and 3 (February-March, 1975).

Baker, R., and R. Herdt. "Who Benefits From the New Technology?" The Rice Economy of Asia. Washington, D.C.: Resources for the Future, 1984.

Barbier, P. "Inverse Relationship Between Farm Size and Land Productivity: A Product of Science or Imagination?" Economic and Political Weekly, Vol. XIX, Nos. 52 and 53 (December 22-29, 1984), pp. A-189 to A-198.

Bardhan, P. K. Land, Labor, and Rural Poverty. New York: Columbia University Press, 1984.

Bates, R. Markets and States in Tropical Africa. Berkeley: University of California Press, 1981.

Benito, C. A. "Rural Development in Puebla: A Minimum Package Approach." University of California, Department of Agricultural and Resource Economics, Berkeley, 1976, 125p.

Berry, A. "Land Reform and the Adequacy of World Food Production," in International Dimensions of Land Reform, J. Montgomery (ed.). Boulder, Colorado: Westview Press, 1984.

Berry, A., and W. Cline. Agrarian Structure and Productivity in Developing Countries. Baltimore: The Johns Hopkins University Press, 1979.

Bhalla, S. "Farm Size, Productivity, and Technical Change in Indian Agriculture," in Agrarian Structure and Productivity in Developing Countries, A. Berry and W. Cline (eds.). Baltimore: The Johns Hopkins University Press, 1979.

Bhalla, G. S., and G. K. Chadha. Green Revolution and the Small Peasant. New Delhi: Concept, 1983.

Binswanger, H., and M. Rosenzweig. "Behavioral and Material Determinants of Production Relations in Agriculture," Journal of Development Studies (April, 1986).

Braverman, A., and T. N. Srinivasan. "Agrarian Reforms in Developing Rural Economies Characterized by Interlinked Credit and Tenancy Markets." World Bank, Staff Working Paper No 433, Washington, D.C., October, 1980.

Burke, R. "Green Revolution Technology and Farm Class in Mexico," Economic Development and Cultural Change, Vol. 28, No. 1 (1979), pp. 135-154.

Chayanov, A.V. "On the Theory of Non-Capitalist Economic Systems," in The Theory of Peasant Economy, D. Thorner et al. (eds.). Homewood, Illinois: Richard D. Irwin, 1966.

Cheung, S. N. The Theory of Share Tenancy. Chicago: University of Chicago Press, 1969.

Commander, S. and P. Peek. "Oil Exports, Agrarian Change, and the Rural Labour Process: The Ecuadorian Sierra in the 1970s." International Labor Office, World Employment Programme Research, Working Paper No. 10-6/WP63, Geneva, November, 1983.

Cornia, G. "Farm Size, Land Yields, and the Agricultural Production Function: An Analysis for Fifteen Developing Countries," World Development, Vol. 13, No. 4, (1985), pp. 513-534.

de Janvry, A. The Agrarian Question and Reformism in Latin America. Baltimore: The Johns Hopkins University Press, 1981.

de Janvry, A. and E. Sadoulet. "Agricultural Price Policy" in General Equilibrium Frameworks: Results and Comparisons. University of California, Department of Agricultural and Resource Economics, Working Paper No. 342, Berkeley, March, 1986.

de Janvry, A. and K. Subbarao. Agricultural Price Policy and Income Distribution in India. Delhi: Oxford University Press (in press).

Deere, C. D. and A. de Janvry. "A Conceptual Framework for the Empirical Analysis of Peasants," American Journal of Agricultural Economics, Vol. 61, No. 4 (November, 1979), pp. 601-611.

Deere, C. D., and R. Wasserstrom. "Ingreso Familiar y Trabajo No Agricola entre los Pequenos Productores de America Latina y El Caribe", in Agricultura de Ladera en America Tropical, A. Novoa and J. Posner (eds.). Costa Rica, Turrialba: CATIE, 1981.

Feder, G. "The Relation Between Farm Size and Farm Productivity," Journal of Development Economics, Vol. 18 (1985), pp. 297-313.

Feder, G., R. Just, and D. Zilberman. "Adoption of Agricultural Innovations in Developing Countries: A Survey," Economic Development and Cultural Change, Vol. 33, No. 2 (January, 1985), pp. 255-298.

Just, R. E., and D. Zilberman. "Stochastic Structure, Farm Size, and Technology Adoption in Developing Agriculture," Oxford Economic Papers, Vol. 35 (1983), pp. 307-328.

Lehmann, D. "The Death of Land Reform: A Polemic," World Development, Vol. 6, No. 3 (1978), pp. 339-345.

Lin, W. G. Coffman, and J. B. Penn. "U.S. Farm Numbers, Sizes, and Related Structural Dimensions: Projections to Year 2000." U. S. Department of Agriculture, Economics, Statistics, and Cooperative Service, Technical Bulletin No. 1625, July, 1980.

Lipton, M. "Land Assets and Rural Poverty." World Bank, Staff Working Paper No. 744, Washington, D. C., August, 1985.

Lipton, M., and R. Longhurst. "Modern Varieties, International Agricultural Research, and the Poor." Consultative Group on International Agricultural Research, Study Paper No 2, Washington, D. C., 1985.

Monardes, A. "An Econometric Model of Employment in Small Farm Agriculture: The Central Valley of Chile." Cornell University, Latin American Studies Program, Dissertation Series No. 78, Ithaca, New York, 1978.

Roy, P. "Transition in Agriculture: Empirical Studies and Results," Journal of Peasant Studies, Vol. 8, No. 2 (1981).

Roemer, J. A General Theory of Exploitation and Class. Cambridge: Harvard University Press, 1982.

Ruttan, V. "Assistance to Expand Agricultural Production," World Development, Vol. 14, No. 1 (1986), pp. 39-63.

Singh, I. "Small Farmers and the Landless in South Asia." The World Bank, Employment and Rural Development Division, Development Economics Department, Washington, D.C., July, 1982.

Todaro, M. "A Model of Labor Migration and Urban Unemployment in Less Developed Countries," American Economic Review, Vol. 59, No. 1 (March, 1969), pp. 138-148.

Warriner, D. Land Reform in Principle and Practice, Oxford: Clarendon Press (1969).

APPENDIX

The relations between land productivity and farm size in Figure 1 can be derived from the following model of household decisionmaking. This model is a modification of those developed by Bardhan and by Feder.

The objective of the household is to maximize $U(Y, \ell_1, \ell_2)$

where

$$Y = Q[\alpha_A A, \alpha_L(\bar{e}L_1 + \bar{e}L_2 + eN)] + W(H - N) - R(A - \bar{A}) - K_0$$

subject to

working capital constraint,

$$WN + K_0 + R(A - \bar{A}) + \alpha_A + \alpha_L = B + WH = K$$

time constraint for

captive labor, $L_1 + \ell_1 = \bar{L}_1$

tradable labor, $L_2 + S + H + \ell_2 = \bar{L}_2$

where

supervision function, $e = \bar{e} - e(\frac{S}{N})$ $\quad e \leq \bar{e}, e' > 0$

embodied technological change,

landsaving, $\alpha_A = \alpha_A(K)$ $\quad \alpha_A' > 0$

laborsaving, $\alpha_L = \alpha_L(K)$ $\quad \alpha_B' > 0$

credit constraint, $B = a\bar{A}^{\varepsilon}$ $\quad \varepsilon \geq 1$

limit of captive labor use, $L_1 = 0$ if $L_2 = S$.

The definitions are:

$U(\cdot)$ = utility for leisure
Y = household income
$Q[\cdot]$ = farm production function with output Q
$\underline{A}$ = size of operational unit
A = size of ownership unit
α_A, α_L = embodied technological change coefficients for landsaving and laborsaving capital, respectively
$\underline{e}$ = efficiency of hired labor, $e \leq e$
$\bar{e}$ = efficiency of family labor
$\bar{L}_1$ = total captive family labor time
L_1 = captive family labor used in farm production
ℓ_1 = leisure time for captive family labor
$\bar{L}_2$ = total tradable family labor time
L_2 = tradable family labor used in farm production
ℓ_2 = leisure time for tradable family labor
S = tradable family labor used for supervision of hired labor
H = tradable family labor hired out
N = labor hired in
W = wage
R = land rent
K_0 = capital starting cost
B = credit
e = elasticity of access to credit based on A as collateral
K = working capital
e_0 = transaction cost in hiring labor (search cost)

As farm size increases from zero (landless) to the largest farm size (capitalist farmers), the following utility-maximizing labor deployment strategies define the following social classes (Roemer; Bardhan):

Household type	Size of ownership unit A	Family labor Captive L_1	Tradable L_2	S	H	Hired labor N
Landless	$0-A_1$	0	0	0	+	0
Subfamily farm	A_1-A_2	+	+	0	+	0
Family farm	A_2-A_3	+	+	0	0	0
Rich farmer	A_3-A_4	+	+	+	0	+
Capitalist farmer	A_4-A_{max}	0	0	+	0	+

APPENDIX TABLE 1

Relation Between Land Productivity and Farm Size

Country	A_F	a_0	a_1 $A < A_F$	b_1 $A \geq A_F$	R^2
	hectares				
Sudan	2.25	175.3 (7.07)	-46.6 (-3.45)	-2.53 (-1.45)	.67
Syria	3.50	264.1 (4.93)	-72.2 (-2.96)	-.16 (-.11)	.45
Ethiopia	1.75	163.3 (4.27)	-44.3 (-1.69)	-16.0 (-2.43)	.57
Nigeria	1.75	743.9 (14.1)	-189.8 (-5.51)	-13.1 (-4.62)	.87
Tanzania	1.75	488.5 (9.96)	-237.6 (-6.13)	0.76 (0.21)	.78
Uganda	3.50	304.1 (12.11)	-52.3 (-3.06)	-43.4 (-6.36)	.90
Barbados	1.50	4,388.3 (2.71)	-3,588.0 (-1.04)	113.7 (0.06)	.15
Mexico	5.00	281.0 (0.94)	-1.07 (-0.01)	-6.17 (-0.85)	.09
Peru	4.50	84.8 (0.34)	81.9 (1.01)	-1.18 (-.06)	.13
Bangladesh	1.25	568.7 (4.11)	-36.8 (-.20)	-27.9 (-.40)	.03
Burma	1.75	813.5 (14.8)	-365.2 (-9.74)	-2.11 (-.70)	.93
India	1.75	684.5 (5.27)	-21.0 (-.25)	-43.7 (-3.11)	.61
Nepal	1.75	747.9 (12.3)	-237.7 (-4.48)	-1.68 (-.07)	.77
Korea	3.50	1,045.2 (5.26)	-194.2 (-2.75)	-6.58 (-.34)	.73
Thailand	1.50	164.8 (4.52)	-23.9 (-.78)	-7.8 (-1.07)	.76

NOTE: The estimated equation is:

$$\frac{GO}{LUI} = a_0 + a_1(A \times LUI) \quad \text{for} \quad A < A_F$$

$$\frac{GO}{LUI} = b_0 + b_1(A \times LUI) \quad \text{for} \quad A \geq A_F$$

subject to:

$$a_0 + a_1(A_F \times LUI) = b_0 + b_1(A_F \times LUI)$$

where

GO = gross output per hectare in 1970 dollars (U. S.)
LUI = land-use intensity ≤ 1
A = farm size in hectares
A_F = size of a family farm.

and figures in parentheses are t ratios.

APPENDIX TABLE 2

Farm Size and Sources and Levels of Income

Cajamarca (Peru), Garcia Rovira (Colombia), and South Bolivia

Country and farm size	Year	Share of farm households	Share of income derived from: Farm activities	Wages	Other activities	Total annual income
hectares		percent				dollars (U. S.)
Cajamarca (Peru)	1973					
0-0.25		13.3	20.3	55.5	24.2	296
0.25-3.50		59.0	24.0	48.6	27.4	206
3.50- 11		16.9	55.4	23.5	21.1	270
11- 30		7.7	82.0	11.4	6.6	582
30-100		3.1	89.6	5.7	4.7	1,320
Total		100.0	46.7	33.8	19.5	293
Garcia Rovira (Colombia)	1972					
0- 4		20	50	46	4	365
4-10		45	77	19	4	543
10-50		28	74	21	5	919
50+		7	98	0	2	4,041
Total		100	72	24	4	1,441
South Boliva	1976-77					
0- 1		19	27	73		309
1- 2		21	40	60		262
2- 5		27	45	55		372
5-10		15	63	37		373
10+		8	67	33		343
Total		100	44	56		327

Puebla (Mexico), 1970

Farm types[a]	Farm size	H/L[b]	N/L[b]	Share of farm households	Share of income derived from: Farm activities	Wages	Other activities	Total annual income
	hectares			percent				dollars (U. S.)
F_1	0-2	0-2	0	20.9	43	57	0	271
F_2	0-4	2+	0	9.3	18	82	0	360
F_3	0-4	0	0	12.6	29	69	2	629
F_4	2-4	0-2	0	27.5	57	43	0	402
F_5	4+	0	0-2	25.3	68	32	0	675
F_6	4+	0	2+	4.4	83	9	8	1,974
Total				100.0				

(Continued on next page.)

Appendix Table 2--continued.

Region IV (Chile), Sierra (Ecuador), Punjab and Harvana (India), and Gujarat and Maharastra (India)

Country and farm size	Year	Share of farm households	Share of income derived from: Farm activities	Wages	Other activities	Total annual income
hectares		percent				dollars (U. S.)
Region IV (Chile)	1976					
0- 0.5[c]		19	20	63[d]	17	618
0.5- 1		21	38	45	17	702
1- 2		19	54	37	9	1,239
2- 5		25	75	21	4	1,941
5-10		11	85	13	2	3,676
10+		4	84	16	0	8,628
Total		100	54	36	10	1,735
Sierra (Ecuador)	1974					
0- 1		34	23	63	14	561
1- 5		43	57	35	8	579
5- 20		16	79	12	9	1,218
20- 50		4	87	6	7	2,022
50-100		2	87	6	7	7,547
100+		1	72	4	24	13,368
Total		100	51	39	10	1,008
Punjab and Harvana (India)	1970-71					
Landless		42	51	15	35	698
0- 1		1	7	0	93	470
1- 2.5		1	63	27	10	379
2.5- 5		6	92	0	8	831
5- 7.5		10	86	1	13	795
7.5-10		12	96	0	4	921
10-15		13	89	0	11	1,060
15-25		12	98	0	2	1,222
25+		4	98	0	2	1,493
Total		100	79	5	16	891
Gujarat and Maharastra (India)	1970-71					
Landless		29	10	50	39	307
0- 1		3	35	49	17	277
1- 2.5		6	46	39	15	235
2.5- 5		10	68	22	10	337
5- 7.5		8	85	11	5	337
7.5-10		9	84	8	9	536
10-15		11	94	4	2	548
15-25		13	96	1	4	1,041
25+		11	100	0	2	746
Total		100	73	16	12	415

(Continued on next page.)

Appendix Table 2--continued.

Kerala and Tamil Nadu (India), and Bengal and Bihar (India)

Country and farm size	Year	Share of farm households	Share of income derived from: Farm activities	Wages	Other activities	Total annual income
hectares		percent				dollars (U. S.)
Kerala and Tamil Nadu (India)	1970-71					
Landless		33	7	33	60	359
0- 1		29	45	11	44	513
1- 2.5		10	56	6	38	549
2.5- 5		11	68	2	29	659
5- 7.5		3	72	1	26	670
7.5-10		3	91	0	8	971
10-15		2	92	0	8	1,443
15-25		2	92	0	7	1,100
25+		1	89	0	10	4,087
Total		94	49	13	38	534
Bengal and Bihar (India)	1970-71					
Landless		36	21	33	46	337
0- 1		7	24	19	55	323
1- 2.5		13	67	9	24	312
2.5- 5		16	65	3	31	441
5- 7.5		9	82	1	18	478
7.5-10		10	79	1	20	534
10-15		7	89	0	11	550
15-25		2	96	0	4	1,043
25+		1	98	0	2	746
Total		100	47	12	30	415

United States, 1974

Sales class	Farms	Net farm income	Off-farm income	Farm program payments	Total net income per farm
	percent	dollars (U. S.)			
0-2.5	31.2	-412	12,411	400	12,399
2.5-5	11.8	-1,039	11,566	715	11,242
5-10	12.0	1,401	9,640	811	11,852
10-20	12.6	4,135	7,444	1,083	12,662
20-40	13.1	9,499	5,512	1,336	16,347
40-100	13.2	20,453	4,997	1,677	27,127
100+	6.2	63,287	8,060	3,890	75,237
All	100	8,890	9,487	1,305	19,682

[a] F_1 = very poor peasants; F_2 = proletarianized peasants; F_3 = prosperous peasants; F_4 = poor peasants; F_5 = family farms; and F_6 = business farms.

[b] H = family labor hired out
N = family labor hired out
L = family labor used in home production.

[c] Basic irrigated hectares.

[d] Off-farm income which is mainly wages.

Sources: For Cajamarca (Peru), see C. D. Deere and A. de Janvry; for Garcia Rovira (Colombia), A. de Janvry (1981a); for South Bolivia, C. D. Deere and R. Wasserstrom; for Puebla (Mexico), C. A. Benito; for Region IV (Chile), A. Monardes; for Sierra (Ecuador), S. Commander and P. Peek; for India, I. Singh; and for the United States, W. Lin, G. Coffman, and J. B. Penn.

PART TWO

The Domestic Perspective

9

The Current Setting of U.S. Agriculture

D. Gale Johnson

For those who have been saying that U. S. agriculture would benefit from a decline in the foreign exchange value of the dollar, from establishing price supports at more realistic levels, and from a fall in real and nominal interest rates, it is now possible to say that we can see the light at the end of the tunnel. The current economic and financial circumstances of U. S. farming are about as bad as they are going to get. We may see some further moderate declines in farm land values in some states, though the substantial commitment of crop subsidies for the next five years should prevent sharp declines.

But the relevant question is: "How long is the tunnel?" There are now too many resources engaged in U. S. agriculture. Our capacity to produce farm products is substantially greater than the amount demanded at prices that will provide returns to agricultural resources equal to what comparable resources receive elsewhere in the economy. The magnitude of the disequilibrium--the excess capacity that we have created--is such that even $55 billion to $60 billion in governmental expenditures upon farm income and price programs over the next three years will not be sufficient to prevent some further decline in asset values and significant further shrinkage in the real capital employed in agriculture.

Unfortunately, the Food Security Act of 1985 contributes little to the resource adjustment process other than buying some time to correct the consequences of serious policy errors committed over the past decade. Other than permitting market prices to decline to levels that may permit the market to clear, there is little that is market-oriented in this legislation. In fact, given the enormous subsidy cost and the market manipulations, such as forgiving part of price support loans and various forms of export subsidies, this may be the least market-oriented farm legislation of the past two decades. The hopes of many of us that the 1985 farm legislation would take a significant turn toward greater market-orientation have been dashed.

How far this farm legislation has departed from any touch with the reality of markets is made starkly clear by a headline in a leaflet of the University of Georgia agricultural extension service: FARM PROGRAMS PROVIDE BEST MARKET IN 1986. This puts the matter about as succinctly as one can.

I shall emphasize three aspects of the current setting of U. S. agriculture. The first is a review of some of the features of the 1985 legislation, some of which are really quite remarkable. The second is a brief review of the international market situation and the prospects for regaining a significant fraction of our lost export markets. Finally, I shall consider how the 1985 farm legislation affects the prospects for success of prospective negotiations on agricultural trade in GATT.

But before we turn to these three points, there are two other topics that merit our attention. Both involve looks to the past, one for four decades and the other for just the past five years.

Some Wisdom From the Past

Before we try to describe and understand the current setting of U. S. agriculture, it may be useful to pause for a brief historical note that may help us understand how the agricultural problems of recent years have been created. It isn't as though we haven't been here before. We have. And it wasn't so very long ago.

During World War II U. S. agriculture was urged to expand its productive capacity to help prevent food shortages and even starvation during and after the war. But as the war came to an end, it was clear to some that U. S. agriculture's capacity to produce was greater than the market could absorb. Consequently at the time, we had several organizations calling for the establishment of market-oriented farm programs—farmers should produce what the market wanted at prices that equated supply and demand and should not be encouraged by governmental programs to produce more than these quantities. Consider the following:

> "The technological revolution in agriculture has made it possible for the farm family to handle larger farms, increase output per worker greatly, and reduce costs. This trend will continue in the future. For many commodities it has increased supply more rapidly than demand in the United States, and this has resulted in a long-run downward pressure on agricultural prices. An essential need in such cases is to reduce the overpopulation in agricultural areas so that commercial farm families can operate adequate farm units on a profitable basis.
>
> The needed long-run adjustments in American agriculture are not necessarily accomplished by present price-support programs together with systems of production quotas. This may lead in some cases toward encouraging farmers to remain in uneconomic types of production instead of shifting to lines more profitable for them.
>
> Our price policies in the past have had a tendency to price our export commodities out of the world market and have led to the use of export subsidies. Such actions have led to retaliation, international friction, and a strangulation of world trade; this is out of harmony with the need to establish international cooperation and an expansion of world trade as part of the broad program for world peace."

These perceptive words were contained in the report of what was known as the Colmer Committee, or more accurately, the Special Committee on Post-war Economic Policy and Planning and the report was issued in 1946. Every word is applicable to 1986, four decades, several farm bills, and billions of dollars of federal expenditure later. The disequilibrium in agricultural resources is now at least as great as in 1946 and in some cases perhaps even greater. For the past several years our price support policies have priced our exports out of the world market. And while we have complained to the European Community about their export subsidies, we have a target price system that provides large subsidies on our exports of wheat, feed grains, cotton, and rice. And, we now have to face the fact that instead of reducing such subsidies, the 1985 farm bill increases them substantially.

How We Got to Where We Are: The 1981 Farm Bill

I have elsewhere argued that the 1981 Act was a careless piece of legislation (1984a). I said this in the context of the high budgetary costs, which have been several times those anticipated, of the loss of international markets because the price supports made us the residual supplier, and because of the legislative failure to permit any significant degree of flexibility to respond to changing conditions, such as the sharp increase in the foreign exchange value of the dollar. The blame, I believe, was upon the Administration as well as the Congress. Some have claimed that the legislation was not carelessly drawn but that there were those in Congress who well understood what they were doing. The high target and support prices were expected to result in substantial deficiency payments and high governmental costs unless the international and domestic markets for farm products were very strong. If the markets were not strong, the appropriate response was intended to be supply management and output reduction.

Thus, it could be argued, the high target and support prices were not set in error and, in fact, had the intended consequence. If accurately described, this policy position follows from a lack of confidence in the ability of the American farmer to respond to changes in economic conditions. It also assumes that the measures the government can take will have desirable consequences for farmers. The validity of this description of Congressional intent is supported by the minimal changes that Congress has been willing to make in target and support prices even when the evidence is at hand that the costs of continuing with the mandated prices would be enormous and the benefits very small.

The 1981 Act left American agriculture in much greater financial difficulty than when the act was passed. Of course, not all of the factors that affected agriculture adversely between 1982 and 1985 were due to the provisions of the 1981 act. Many had origins in the late 1970s--low and often negative real interest rates encouraged borrowing and higher farm land prices, low foreign exchange value of the dollar resulted in a high rate of growth of exports, and a loose monetary policy was followed by high rates of inflation and high nominal rates of interest. These and other factors, including the debt problems of many of our export customers and the sharp increase in the foreign exchange value of the dollar, combined with the inflexible price supports imposed by the legislation caused a sharp decline in our farm exports. The 1981 act was created on the basis of the false perception that world demand for food would require the full productivity capacity of world agriculture to meet that demand. William Lesher, former Assistant Secretary of Agriculture, in 1984 described the prevailing views as follows:

> "At the time the 1981 Farm Bill was formulated, the main concern was that world food needs would outpace production. Many believed--inside and outside the government--that the agricultural issue of the 1980's was going to be how to produce enough for a starving world rather than surpluses. Many also believed that the United States was the only country that possessed the potential to expand food production enough to meet world needs."

As late as mid-1980, the Secretary of Agriculture (Bergland) would state:

> "The era of chronic overproduction, surplus disposal problems and a seemingly infinite supply of resources is over. We have moved into a new era--one in which food supplies are tighter, more food is being consumed, and the resources which produce that food are becoming depleted."

In the same speech, the Secretary repeated a statement made by another: "Some gloomy predictions have American food exports ending by the year 2000 as a result of environmental, energy and economic constraints."

In this setting of gross misinformation, Congress felt itself capable of establishing both target and support prices for grains and cotton and price supports for sugar and soybeans and honey for 1982 through 1985. And the Administration was willing to accept what Congress offered it, apparently in exchange for support for tax reduction. Memories were apparently very short. The more reasonable levels of price supports and target prices in 1977 and 1978 had led to significant excess production, even with the low and declining foreign exchange value of the dollar. In terms of 1985 prices, the costs of farm price and income support programs in 1978 was about $10.5 billion--exceeded since then only by the 1983 extravaganza. Of course, 1986 will clearly set a new record. But as I look back, I find it curiouser and curiouser why all of us, and I include the economists in this as well as the politicians, did not at least pause to consider the longer term implications of the excess productive capacity displayed in 1977 and 1978. But hardly any one did.

Apparently we were lulled into a false sense of prosperity by the return of feed use of grains to the levels of the early 1970s, following the sharp decline that occurred in 1974/75. The growth in domestic feed use came at the same time as our grain exports increased by a third. What we failed to see was that we supplied a large share of the 50-million-ton increase in world exports of wheat and feed grain from 1975/76-1977/78 to 1980/81-1981/82. We supplied about 60 percent. In turn, the centrally planned economies accounted for slightly more than 60 percent of the increase in world grain imports. In 1969-71 the U. S. had about a third of world trade in grains. By 1980-82 our share had increased to 48 percent. Somehow we were able to ignore the highly probable, namely that the grain imports of the centrally planned economies could not continue to increase at an annual rate of 25 percent as occurred between 1969-71 and 1980-82, and that our share of world grain trade might decline to nearer the percentage that we had before the period of rapid growth of world trade.

Major Features of 1985 Farm Bill

It is possible to present only some of the features of the very complex 1985 farm legislation. I shall concentrate on those that relate to the structure of incentives as relevant to assisting agricultural adjustment and those that may influence agricultural exports and our position as a traditional supporter of liberal trade in farm products.

Some major features of the legislation are well known. Target prices were held at the 1985 levels for two years and were then to be reduced very slowly over the next three years. Price supports for grains were significantly reduced starting with 1986 and with the use of some magic, even for 1985 for rice and, to some degree, for cotton.

The changes in price supports for the grains are substantial, especially for wheat and the feed grains. To record what I am sure you already know, let us compare the price support loan rates for 1985 and 1986, with 1985 coming first: wheat, \$3.30 to \$2.40; corn, \$2.55 to \$1.92; rice, \$8.00 to \$7.20; cotton, \$0.575 to \$0.55. For wheat and corn the reductions in loan rates are about a fourth, the reduction for rice is a tenth, which the change for cotton was minimal. Price support levels were not significantly reduced for sugar, peanuts, dairy products, or soybeans for the immediate future. However, reductions in dairy and soybean price supports can be made starting in 1988. In fact, in nominal terms the price support for peanuts is to be increased based on production cost increases from 1981 to 1985. The soybean price support was not changed, even though our soybean stocks have increased to relatively high levels due to the market price-loan rate relationships of the past two years.

One way of describing the degree of market-orientation of our farm legislation is to estimate the shares of the returns for a crop that come from the market and the government. This is not a totally accurate measure since the amount that comes from the market may reflect significant governmental intervention, though this may not be too significant for the next two or three years. However, as I present such measures it will be evident that it is unnecessary to engage in a refined analysis to show the substantial change between the 1981 and 1985 farm bills in the degree of market-orientation as related to the production incentives provided to farmers.

One such indicator is the change in the absolute size of deficiency payments between 1985 and 1986. For wheat, the maximum deficiency payments will increase from $1.08 per bushel to $1.98; for corn, from a maximum of $0.53 to $1.11 per bushel, and from $2.90 to $3.70 for a hundredweight of rice. The deficiency payment for cotton will increase hardly at all, since the price support for cotton declines but 2.5 cents per pound, but, as in the case of rice, the full deficiency payment will increase radically. Thus the deficiency payments as officially described can amount to the following percentages of the prices that farmers will received during the first few months of the marketing year, as seems highly probable: wheat, 82.5 percent; corn, 58 percent; and rice, 50 percent. For wheat and corn these will represent substantial changes from the year before. All government payments, not just deficiency payments, for wheat in 1985 were equal to 34 percent of the market value of wheat production, 12 percent of the value of corn, and 25 percent for cotton.

There is a startling new provision in the 1985 legislation. This provision has the potential for making a mockery of the price support system, turning it into a thinly disguised price subsidy similar to the deficiency payments, but not to be counted toward the $50,000-per-person ceiling on deficiency payments. It moves the United States into the category of a number of developing nations that make loans to farmers with no intention of collecting all or a large part of the loans.

The first use of this provision has already occurred and it was applied not to the 1986 production but to 1985 price support loans. As of April 15, 1986, rice farmers could repay their price support loans at less than the original loan amount and receive their rice in return. The payment was made at what the Department of Agriculture determined was the world market price, a somewhat elusive concept, given the wide price variations for rice. As I will note later, this action has been strongly protested by the Government of Thailand as disruptive and predatory and distinctly unfriendly. The impact on the market price of rice in the U. S. and on our rice in world markets was abrupt and substantial. As of April 15, Arkansas number 2 milled rice was priced at 16.5 to 18.0 per pound; by May 1 the price had fallen some 22 percent to a range of 13.0 to 14.0. If the rice price stays in this range, the deficiency payment plus the amount of the repayment subsidy for 1986 will mean that rice producers will receive as much per hundredweight in federal subsidies as from the market for the 1986 rice crop. If output is held to the low 1984 level of about 100 million hundredweight, the two payments will cost about $560 million. This is for the benefit of fewer than 11,000 producers--an average cost per producer of more than $50,000. Perhaps the 400 largest rice producers will receive total subsidies in excess of $100,000 from the two subsidies.

The second use of a similar provision will be for cotton as a payment of the difference between the 1985/86 loan rate of 57.3 per pound plus carrying charges and the world price on all free stocks of cotton held on August 1, 1986. Free stocks of cotton are all stocks in the United States not held by the Commodity Credit Corporation. Thus every individual or corporation owning any cotton on that date will receive a negotiable certificate with a value equal to the product of the amount of cotton owned and the subsidy per pound as determined by the USDA. The reason for this scheme is to encourage people to redeem 1985/86 price support loans in order to create the free stocks and get the subsidy and thus lower the market price of the cotton. Nothing has been said about the decline in value of textiles owned by mills and wholesalers when the price of cotton drops precipitously in early August. Perhaps there will be a subsidy for them as well.

Similar loan payment provisions may apply to wheat, feed grains, and soybeans for 1986 to 1990 crops. At the discretion of the Secretary of Agriculture, loan repayment can be made at rates that are as much as 30 percent below the basic loan rate. This means that the potential degree of subsidy for wheat and feed grains can be substantially greater than estimated on the basis of the deficiency payments alone. If this provision is used in an effort to sharply reduce CCC stocks, the farm price of corn could be reduced to $1.35 per bushel, a mere 43¢ in 1967 prices. I am not predicting that this is what will happen, but the illustration indicates that risk and uncertainty in the grain markets may have been substantially enhanced by legislation with the title of Food Security.

Some part of the payments for diversion and the advanced deficiency payments (40 percent of the estimated total) will be paid in kind and are not included in the $50,000 payment limitation. Actually the payment is in kind only if the recipient of certificate has a CCC loan at the time of signup. If the farmer has no such loans, the "generic certificates" are to be dollar denominated, negotiable and payable to the bearer in kind. For the original holder they are equivalent to cash since they can be sold for cash, but yet their value is outside the payment limitation.

The 1985 legislation does provide for acreage reductions of 12.5-17.5 percent for feed grains, 15-22.5 percent for wheat, up to 25 percent for cotton, and up to 35 percent for rice. These are the potential ranges for 1986; slightly different ranges apply to other years for wheat and feed grain. There is also provision for a very modest amount of additional paid diversion. Another new wrinkle is an underplanting provision that permits producers to receive deficiency payment protection on 92 percent of their permitted plantings if they plant at least 50 percent of their permitted acreage to the program crop and the remaining acreage to a conservation use or a nonprogram crop other than soybeans or extra-long staple corn. It will be interesting to see if the potential of substantial further subsidies achieved by reducing the repayment rate for price support loans means that this provision will be little used. Or if it is used, what crop's profitability will be destroyed by being planted on such areas? Will it be sunflowers? Or flaxseed? Or will it end up as additional high quality pasture and add to the supply of beef calves? Beef cattle producers are among the principal victims of the dairy herd reduction program, so perhaps they will receive a double whammy from the 1985 farm bill.

In judging what the 1985 farm legislation may accomplish, we do have to recognize that the legacy left from the 1981 farm bill leaves a great deal to be desired. Large numbers of farmers face significant financial difficulties, and large stocks of the grains, soybeans, and cotton now exist, primarily as a result of the inflexible provisions of the 1981 farm bill.

At the beginning of the 1981/82 crop year the major wheat exporters held stocks of wheat in the amount of 45.5 million metric tons, of which the United States held 26.9 million tons. At the end of the 1985/86 year, stocks of the major exporters are projected to be 81 million metric tons, of which the U. S. will hold 49 million tons. For feed grains, the major exporters had stocks of 46 million tons at the beginning of the 1981/82 crop year, of which the U. S. held 35 million tons. At the end of the 1985/86 year, the projected stocks of the major exporters will be 118 million tons, of which we are the proud owners of 109 million tons.

The 1985 legislation starts with a massive negative legacy from the 1981 legislation--extremely large stocks of grains, soybeans, and cotton. These stocks will, for the rest of the 1980s, clearly minimize any positive market price effects that might otherwise have followed from the decline in the foreign exchange value of the dollar and the increase in import demand for farm products that will occur as the debt crisis of several developing countries diminishes.

Regaining Export Markets

I shall now consider how effective the legislation will be in increasing exports of farm products. The reduction in price support levels, including the provision for the partial forgiveness of CCC loans, should stop the erosion in our export shares in Fiscal Year 87. However, the short run effect has been negative, since we are saying to potential buyers--wait a few months and you can get it cheaper. Surely some of the lag in Fiscal Year 86 exports is the result of the price provisions of the 1985 farm bill.

Ignoring the short run negative impact upon exports, will the decline in price supports do more than stop the erosion of exports, either in terms of absolute amounts or export share? Our share of world exports of wheat and feed grains in 1985 was 41 percent, down from 57 percent for 1972-83. Even if our share of world grain exports gradually increases to 50 percent, the probable level of world grain trade for the rest of this decade would mean that our grain exports would remain at or below our levels of 1979-81. The reduction in grain price supports will help us to gradually increase our share of world grain trade, though it is not clear that our soybean exports in 1986 and 1987 will be facilitated by the legislation.

How much increase in our share can we expect in two or three years? Even with the decline in the foreign exchange value of the dollar, our major competitors are unlikely to make substantial adjustments in their production plans in that period of time. Canadian wheat producers will sow more wheat in 1986 than in 1985 if the early 1986 survey of planting intentions reflects actual decisions. There is no evidence that in a period of two or three years, there will be substantial changes in grain production in the European Community; thus EC grain exports are unlikely to decline. Argentina may produce less grain over the next two or three years, but not because of our changes in price supports but because the Argentine government is once again imposing heavy export taxes on most agricultural exports and thus reducing incentives to produce.

Consequently if total world grain and cotton trade increase only moderately in response to declining market prices, we are unlikely to see any significant increase in either our share of world grain trade or in the volume of our grain, feed, and cotton exports. I am not arguing that there will be no increase in share and that there will be no gain in volume. What I am saying is that any expectation that there will be a major change in our exports of farm products before the end of the decade is almost certain to be disappointed.

The significant lowering of price supports for grains and cotton will help to increase export quantities and to move some of the large stocks. But if we expect that the stocks will disappear in a year or two, we are only fooling ourselves. Assume that over three years our exports of feed grain were to return to the 1980/81 level of nearly 70 million tons from the 1985/86 estimated level of 50 million tons and that during each of the three years current production equals domestic use plus 50 million tons. Our stocks of feed grain would decline to just a little less than 70 million tons or about 30 percent of a year's production. Even at this level, stocks would act to depress market prices.

If we could regain the 1980/81 level of wheat exports in three years, this would reduce our stocks to reasonable levels in that time, if our production were to average 65 million tons (2.4 billion bushels) during the period. But even with the sharp cut in the wheat price support, there can be no reasonable expectation that our exports can return to the 1980/81 level of 40 million tons, let alone the higher levels of the next two years.

World grain imports are down and are likely to remain at recent levels or even decline over the next few years. The Chinese were large importers of grain in the early 1980s; they will import little grain in 1986/87 and for the rest of the decade. We can't count on significant increases in imports by the USSR; with the decline in foreign exchange earnings due to the sharp fall in oil and natural gas prices, Soviet imports are likely to decline. In fact, Soviet grain imports probably peaked at 54 million tons in 1984/85. In 1985/86, their imports were down substantially to 33 million tons.

Unless the USSR suffers a series of major grain crop failures (crops of less than 160 million tons), there is no basis for assuming any growth in Soviet grain imports for the next several years. There are two significant reasons for this. One is that after decades of neglect, the Soviet farm system now seems to have given forage crops a reasonable priority and the supply of nongrain feeding materials has increased enough in recent years to permit a modest increase in total meat, milk, and egg production form an approximately constant use of concentrates. There remains considerable further room for improvement in this area (Johnson and Brooks). The second reason is the disaster that the Soviet Union has suffered in the sharp fall in the price of oil. In recent years oil has accounted for more than 70 percent of its hard currency foreign exchange and roughly a third of those earnings has been used to purchase food from hard currency areas. The most likely expectation is that Soviet grain and feed imports will decline for the next several years, even at the cost of some reduction in livestock output.

At the end of the 1970s, Eastern Europe was a major grain importer, with net imports of 11 to 15 million tons. As a result of the credit crunch facing all of the countries—especially Poland—and some improvement in agricultural performance, net grain imports in 1982/83 through 1985/86 were just 2.5 million tons annually.

There was a sharp decline in per-capita meat consumption in Poland and a slowing of the growth in the other countries of the region contributing to a reduction in demand for feed materials.

China contributed importantly to the increase in world grain imports in the late 1970s and early 1980s, with net grain imports increasing from 3 million tons to 15 million tons in 1982/83. But in 1985/86 China's net grain imports declined to the 3-million-ton level. The rural reforms in China have been remarkably successful, resulting in an increase in farm output of 50 percent between 1978 and 1984. Some argue that increasing real per-capita incomes will increase the demand for livestock and poultry and thus for feed grains, outstripping the growth in feed supplies. This may happen, but not until well into the 1990s and too late to have any effect on world grain import demand during the life of the Food Security Act of 1985. China has been far more successful in reforming its farm policies than we have. In fact, it may be no exaggeration to say that Chinese agricultural policy is much more market-oriented than is ours. Perhaps we should send a Congressional delegation to Beijing to see how it has been done.

I fear that in recent years there has been too much emphasis upon the effects of our price support levels, the subsidies to exports by the EC, and the high exchange value of the dollar as the causes of the decline in our export position. Thus many people in Washington and elsewhere hold unrealistic expectations concerning the likely growth of our exports now that price supports have been lowered and the foreign exchange value of the dollar has declined. Too many seem to believe that if we only engage in large enough export subsidies, we can regain most of the exports we have lost to the EC and other competitors. Thus we find precipitous and unwise action being taken out of frustration resulting from the failure to realize the unrealistic expectations about our export prospects.

I noted earlier that FY 86 exports have been adversely affected by the expected decline in the market prices of major export products. While the aggregate quantities of agricultural exports may well increase in FY 87, it is not obvious that the value of exports will start to recover until FY 88 and later. The reason is that the price declines for wheat, corn, and cotton will be substantial and the short run elasticity of export demand must be greater than one just to maintain the value of FY

87 exports at the low FY 86 level. It is important that there be a careful analysis of the effects of the lower price support and the decline in the foreign exchange value of the dollar upon our export performance. Only if such analysis is made and accepted can we have any chance of avoiding even further pressure on the USDA to extend additional export subsidies.

The U. S. and GATT Negotiations

The current U. S. farm programs represent a major barrier to successful GATT negotiations to reduce barriers to agricultural trade. In fact, the current legislation may be less congenial to liberal trade than legislation at any time since the mid-1950s. It is hard to see how any negotiator, even Clayton Yeutter, can enter GATT negotiations with any hope of a successful outcome when he has to carry such baggage with him as are found in the Food Security Act of 1985 and our retention of the waiver from Article XI of GATT.

In 1955 we obtained a waiver of indefinite duration that permits us to ignore the provisions, which we wrote, designed to limit the trade disruptions caused by quantitative trade restrictions, such as import quotas. Perhaps the enormous trade interventions of the 1985 legislations will not make any difference to the negotiations; unless we are willing to give up the GATT waiver, the negotiations have little or no chance of success (Johnson 1984b and Johnson, Hemmi, and Lardinois).

Let us assume that Congress is willing to repeal Section 22 of the Agricultural Adjustment Act of 1933, which would make it possible for the U. S. to negotiate with respect to the gradual elimination of import quotas, including voluntary export agreements such as we have used for beef and veal. The major negotiating issue we would then face is when and how could we reduce the enormous subsidies provided by the current farm legislation. Our opposition to the export subsidies of the EC now have a hollow ring. The enormous cost of the deficiency payments, the provision for forgiving a significant part of the price support loans, and the direct measures that we now have for export subsidies surely put us in the same league as the EC as the world's most prolific subsidizer of exports. Perhaps we are on the way to achieving first place in 1986/87.

There are some differences between the deficiency payments and forgiveness of part of price support loans and the direct export subsidies such as are used by the EC. The U. S. programs lower market prices domestically, permitting our consumers access to food at approximately world market prices. The variable levies and export subsidies used by the EC hold consumer prices substantially above world market levels and thus restrict domestic consumption and add to the supplies available for export. Another difference is that the U. S. does attempt to reduce the output of some farm products, though we don't seem to think it important to adequately document how effective our supply management measures have been. We cannot, for example, state with any degree of certainty that farm output in 1985 was larger or smaller than it would have been if after 1976 there had been no target prices and deficiency payments.

Earlier I noted the heavy blow that our rice price policy has inflicted upon Thailand. Do we no longer consider any foreign policy implications of our agricultural programs? We can't say that we are retaliating against Thai export subsidies. Thailand has, in fact, over the years resorted to taxing rice exports and has thus helped to maintain international rice prices. Now we turn around and hit them with a sledge hammer. And all for increasing our rice exports by at most a million tons. I am ashamed for us. Has any one calculated the harm we will do to cotton producers in low income countries by the sharp fall in cotton prices that we will cause as of August 1? I doubt it very much. Most of our competing cotton producers are found in low income developing countries. Generally speaking, the current state of the world cotton market is largely of our doing through high target and support prices that have encouraged both domestic and foreign production. It would seem that we have some responsibility for providing for a slow and moderate transition from where we are to a more market oriented position. Our motives in all this would be less suspect if we had lowered the target price of cotton from its level of 81 cents per pound, a price more than double world market prices as of mid-1986.

About the only positive thing that can be said about the sugar provisions of the 1985 bill is that they are no worse than before. The only difference is that we are now farther down the road than we were five years ago of becoming sell sufficient in caloric sweeteners. By the expiration of the 1985 act we may have come close to reaching that objective. One can only assume that self sufficiency is our national goal, since that is what we are well on the way to achieving. Again, there is no regard for the effects upon other, more efficient sugar producers, many of whom are very poor.

I fear that over the next year or so our policy posture will deteriorate and be less congenial to successful trade negotiations than it is today. As I noted earlier, in spite of lower price supports and the lower foreign exchange value of the dollar, our export quantities will grow slowly over the next 18 months. The value of the exports may actually fall from the low FY 86 level due to the lower prices not being fully offset by greater volume. In due time, perhaps two years or so, the effects of the lower price supports and the exchange rate will be to increase both quantity and value of exports. But even then, the recovery of export values will be slow.

As I noted earlier, there are many who have held unrealistic expecta- tions concerning the speed with which our exports would turn around and increase. When it is realized that the value of our exports may continue to decline for some time, there will disappointment and an emphasis upon even more aggressive behavior on our part.

The U. S. Department of Agriculture recently revised its projection of the value of agricultural exports for FY 86 downward to $28 billion, some 36 percent below the FY 81 peak. The pressure on the Administration has caused it to become more aggressive in the use of its export subsidy authority. More pressure and more disruptive legislation may follow.

Concluding Comments

The 1985 farm bill repeats most of the mistakes of the 1981 act. Production incentives, in the form of target prices, are much too great. The target prices certainly do not say to the farmers that substantial resource adjustments are required. Presumably the low market prices that will prevail for the rest of this decade will transmit a signal of caution. But which signal will be believed? Given the political propensity to blame our difficulties, including the current farm problems upon almost everything other than our farm policies, the view may be held that the target prices are more indicative of market prices down the road than will be the prices over the next year or two. If low market prices, as claimed, are due to the EC high prices and export subsidies, to Japanese import restraints, and the expansion of farm output in Canada, Australia, Argentina, and Brazil, then after we show them just how big and powerful our treasury is, market prices could increase to or near the target level. Or so some will imply and many will want to believe.

My concern that our commitment to a more market-oriented farm policy and a liberal trade policy will further weaken in the years ahead is supported by the following quotation from a recent USDA periodical, which speaks for itself and provides the last chilling word:

> How competing exporters will react to the U. S. policy changes is unknown. Many countries may match the U. S. price declines in the short run. However, if the United States indicates through 1987/88 program provisions that it is committed to recapturing its dominant role in world markets, competitors may be forced to make substantial changes in their farm policies. Except possibly for the EC, no country has the budget to match the United States in a full-blown price war.

REFERENCES

Bergland, Bob. Remarks prepared for delivery by the Secretary of Agriculture before the Conservation Foundation Conferences, Washington, DC, July 14, 1980. U. S. Department of Agriculture, Speeches and Major Policy Releases, July 14 through July 18, 1980.

Johnson, D. Gale. "United States Agricultural Policy--National and International Contexts," in United States Agricultural Policy 1985 and Beyond. Edited by Jimmye Hillman. Department of Agricultural Economics, University of Arizona, 1984, pp. 155-178. (a)

______. "Domestic Agricultural Policy in an International Environment: Effects of Other Countries' Policies on the United States." American Journal of Agricultural Economics 66, No. 5 (December 1984): 735-744. (b)

Johnson, D. Gale and Karen M. Brooks. Prospects for Soviet Agriculture in the 1980s. Bloomington: Indiana University Press, 1983.

Johnson, D. Gale, Kenzo Hemmi, and Pierre Lardinois. Agricultural Policy in an International Framework. A Report of the Trilateral Commission. New York: New York University Press, 1985.

Lesher, William G. "Future Agricultural Policy--A Challenge for All." In Alternative Agricultural and Food Policies and the 1985 Farm Bill. Edited by Gordon C. Rausser and Kenneth R. Farrell. Berkeley, CA: University of California Press, 1985, pp. 37-73.

Special Committee on Postwar Economic Policy and Planning. Postwar Agricultural Policies. Tenth Report of the House Special Committee on Post-war Economic Policy and Planning, August 6, 1946. 79th Cong., 2d Sess. Washington, DC: U. S. Government Printing Office, 1946.

The University of Georgia College of Agriculture, Cooperative Extension Service. Economic Issues in the 1980s. Vol. 2, No. 3, March 1986.

10

A Coherent Policy for U.S. Agriculture

Gordon C. Rausser and William E. Foster

1. Introduction

If there is anything on which everyone interested in food and farm policy can agree, it is that American agricultural policy is an incoherent mess. The huge assortment of programs under the broad title of U.S. Food Policy frequently contradicts its own purposes, sometimes subtly but often blatantly. The conflicts between programs are even sharper when we look at the conflicts between those most affected by policy--livestock producers against grain farmers; conservationists against those who need to use the land, water, fertilizer, and pesticides; and consumers who want cheaper food against producers who want more profitable returns.

The contradictions among components of food policy do not end on the farm or in the supermarket. Programs meant for the agricultural sector, time and again, have been thwarted by and run counter to other national interests and policies. Indeed, the recent history of high interest rates, a strong dollar, and budgetary excess has led some to suppose that, if we could only correct the national economy, the farm crisis would fade away. This notion, of course, downplays the deeper internal problems in agricultural policy. It also ignores that farmers may sometimes enjoy economic conditions that others find distressing.

If we complain that our agricultural policy is contradictory and confusing, what then would a coherent policy be? First of all, such a policy should be constructed to reach certain nationally recognizable goals. Second, the particulars of the policy--the programs and tactics--should not contradict these goals or each other. Finally, the policy should be able to sustain itself when the world changes, e.g., when there are bad harvests, when we embargo grain to the Soviet Union, or when the Federal Reserve tries to control inflation. In short, a more coherent agricultural policy should be one that is clearly articulated, logically connected, and consistent with other national priorities.

But even if we could attain coherence in agricultural policy, would it be desirable? Under current policy, many politicians and farmers feel helpless. Incoherence demoralizes, leads to uncertainty, promotes inefficiency, and generally diverts farmers' attention from economic management of their resources. More importantly, perhaps, the present incoherent situation exacerbates the conflict between individual goals and society's good. It offers many lucrative and ultimately unproductive opportunities for individual groups and factions to work for what is in their near-term, short-sighted interests and against the general welfare. We may be at the point where the most successful farmer is less adept at farming the land than farming the government, and many politicians and lobbyists may be more successful milking the system than voicing the legitimate concerns of constituents.

Market failures often justify governmental action; but in agricultural policy the government itself is subject to failure, making intervention possibly worse than the problem it means to solve. A more coherent policy would tend to reduce the possibility that programs designed to attain broad social goals will change under pressure to reflect the relative lobbying power of narrowly focused groups.

The foregoing is broad and general--just an outline of the characteristics of a more coherent agricultural policy. In order to suggest some specific design, we examine the present situation. What are the possible objectives of agricultural policy? How do programs specific to the agricultural sector conflict with other social interests and national policies? How well do various programs stand up--how consistent, or robust, are they at meeting our goals-- when the world changes?

We turn first to a discussion of the goals of agricultural policy, including that of avoiding political failure in the implementation of programs. Next we analyze the world in which agricultural policy must act, examining the contradictions between sector-specific programs and other policies. We also investigate some specific commodity settings, the interests of agriculturally related groups other than farmers, and the political dimensions of policy construction and implementation. Before detailing our own proposals, we provide a review of some popularly discussed alternatives for policy. Finally, we present a set of proposed programs that we believe would yield a more coherent policy for U. S. agriculture.

2. Public Policy

2.1. The Goals of Farm Policy

The objectives of governmental intervention in food and agriculture are clearly influenced by the "market failure" or equity problem that is presumed to exist. In the case of domestic U.S. agriculture, the rationales for governmental involvement have been many and varied. As stated in the Food and Agriculture Act of 1981, the general purpose of U.S agricultural policy is "to provide price and income protection for farmers, assure consumers an abundance of food and fiber at reasonable prices, continued food assistance to low-income households, and for other purposes" (U.S. Congress, 1981). Given this general purpose, some have argued that the problem in U.S. agriculture is economically depressed farmers who require income enhancement; others have argued that farmers are in a relatively disadvantaged position in the marketplace and require public support in dealing with concentrated buyers of their products; and still others have argued that U.S. agriculture is faced with a large degree of instability in commodity markets adversely affecting not only farmers but also consumers of food and fiber.

Given recent experience, it seems that the most persuasive rationale for an active agricultural policy is the market failure associated with an intolerable degree of instability or excessive risk and uncertainty. Nevertheless, there are other problems of U.S. agriculture that many presume can be corrected by governmental intervention. Thus, a more comprehensive set of objectives than simply risk reduction must be considered.[1]

2.1.1. Redistribution of Wealth

Traditionally, our society has paid much attention to the ideals of an even distribution of wealth: Equity is good; disparity is bad. The historical "farm problem" has been repeatedly characterized by economists and other interested observers as the disadvantaged economic position of farmers. This notion once served as the single most important justification for state action to redistribute society's wealth to the agricultural sector. Policy analysts usually accepted the goal of redistribution without comment, focusing their study on comparing the various means of attaining this goal. Recent discussion, however, has looked askance at income transfers to farming as a heavily weighted objective of policy.

2.1.2. Risk Reduction

The existence of market failures often rationalizes governmental intervention in the marketplace. In particular, the random character of both commodity prices and production is offered to justify public policies aimed at the agricultural sector. A market failure arises from the inability of farmers to adequately trade their risks to others in the economy and from the divergence of social interests from actual farmer response to uncertainty. The tendency of a free economy to yield results different from the socially optimal has been at least tacitly recognized by policymakers. Farm policies, such as price-stabilization schemes and crop insurance, are designed, in part, to affect directly the ability of the agricultural sector to cope with, and respond to, the capricious nature of its physical and economic environment. One way to measure agricultural risk is income variability.

Farming operations have become increasingly reliant on outside sources for their financing and inputs (both material and labor), adding to their sources of risk and uncertainty. As the nature of agriculture has changed, the exposure to more risks has led to apprehension regarding the cash-flow and debt-asset problems of farmers. The focus of public concern has shifted from agriculture's relative poverty to the difficulties of managing in a risky environment without sufficient means of insurance.

The market failure associated with risk and uncertainty is among the most persuasive rationale for an active farm policy. This, of course, is not a necessary and sufficient condition for government intervention in food and agriculture. A sufficient condition is that the loss of economic efficiency in the case of uncorrected market failure is greater than the loss under the government's remedy.

2.1.3. Preservation of the Family Farm and Traditional Rural Communities

Preservation of the traditional concept of the family farm has been a social goal since the beginning of public policies for agriculture. Economically, many small family farms are thought to assure adequate food supplies without domination by a few powerful interests. Politically, family farms are often considered an integral part of the ideal Jeffersonian democracy. The trend of agricultural structure, however, is toward fewer and larger farming enterprises run not so much as a way of life, but as commercial operations, as in other industries.

The relevance of this goal to maintain farms as small family operations will be increasingly questioned as agriculture continues to be further meshed with the rest of the economy. Future farm policy must answer the question: What does society want for the structure of the agricultural industry? This is a question for the long term; short-run policies, meant to respond to immediate difficulties besetting farmers, do not address this issue. Indeed, the concentration on a series of governmental reactions to serious but short-term problems may be taking agriculture further from the basic objective of protecting family farms.

Preserving family farms is but part of the concern. In general, distress on the farm translates into distress in rural banks, suppliers, and other business that depends on the economic health of agriculture. Moreover, as farms grow larger and become more capital intensive and specialized, rural communities suffer. Society puts the goal of maintaining family farms in this larger context of protecting the rural landscape, both physical and social.

2.1.4. Flexible Agricultural Sector

An often ignored but crucial objective of policy is to ensure the ability of the agricultural sector to adapt to changing economic conditions. This is related somewhat to the issues of risk insurance and food security, but another dimension to policymaking may be the desire to avoid future farm policy crises associated with inadequate or mistaken government actions. When production and supply decisions in this country are divorced from the underlying market forces determining demand and production in the rest of the world, conditions arise to pressure U.S. agriculture to become more responsive to the marketplace. Eventually, the market asserts itself in one fashion or another--sometimes with spectacular results.

2.1.5. Conservation of Resources

Resource and environmental issues are traditionally given scant attention in discussions of agricultural policy. The future productivity of U. S. agriculture is at stake; it is a matter of long-term concern. There are two primary reasons that resource management is a matter of public policy. First, society's goals for intergenerational equity may be inconsistent with the short-term objectives of individual farmers, making some form of intervention necessary to preserve adequate and reasonably priced food for the future. Farmers' interests may diverge from society's interests because income and credit constraints make erosive land use and increased water consumption immediately profitable, or because farmers' attitudes about future generations of consumers may simply be contrary to society's view.

Second, institutional arrangements allow the costs of pollution (from sedimentation and farm chemicals) and water consumption to be diffused over many persons and not concentrated among farmers. The rapid depletion of water from the Ogallala aquifer, for example, is a common property problem. Since the water one farmer does not use is available for others, there is no incentive for any individual to husband the resource for its optimal benefit to all. Such externalities result, in part, from the lack of resource property rights. The government may deem water management programs a social good, perhaps limiting water demand by limiting production.

2.1.6. Food Security and Reasonable Consumer Prices

There is a certain public good aspect to the federal government's participation in commodity storage. Society deems important the preservation of adequate food reserves which the private market may be unwilling to guarantee. First, public welfare may be enhanced not only by assurances of supply for this country but by the availability of grain stocks in the event of foreign crop failures, wars, and other catastrophes. Second, in the case of food price inflation, such stocks provide insurance to U.S. consumers that rapid increases in food prices can be moderated.

Typically, governmental programs rely on the manipulation of stocks to affect the price of commodities, but there is perhaps an unintended provision of food security when grain storage exceeds what would otherwise result. When debating and structuring farm policy, some appraisal must be made of the socially optimal level of stored grain; and alternative programs should be judged, in part, by their attention to this question of safe stock levels.

2.1.7. Minimizing Treasury Costs

An additional social goal to be considered in the construction of farm policy is the minimization of Treasury outlays. Many of the objectives discussed above can be reached, given sufficiently large expenditures of tax dollars. There is, of course, a political limit to the amount of funds that can be spent on addressing agricultural issues. Farm legislation is based on the support of the nonfarm population whose acquiescence depends on the financial costs of programs.

2.1.8. Political Failure

In addition to solving market failure, agricultural policy must mitigate the effects of political failure. As we have outlined above, political failure is the tendency of the legislative process to produce policies that do not lead to socially superior outcomes. Political markets induce politicians to consider personal, not public, benefits and costs. As noted above, the existence of market failure is a necessary but not sufficient condition for government intervention.

Political failure has two important effects. First, a policy may be selected that does not solve market problems in an efficient manner but contributes to the short-run goals of politicians. This is the most obvious result of political failure--failure in choice. Economists can do very little to solve this problem other than to inform the public and politicians about available policy choices. The second result, failure in implementation of a policy, is more subtle. Over time, policies may be modified to serve political concerns. Policy analysts should recognize this potential, as well as the additional costs of political failure in implementation, and design proposals that both alleviate market failures and mitigate or avoid political failures.

The idea of political failure is not just economists' response when politicians ignore what theorists recommend. While the term political failure may be basically an academic one, what it describes is a widely recognized phenomenon with a variety of names: pork barreling, cloakroom lobbying, mutual back scratching, and, to much of the public, politics as usual. Not surprisingly, politicians recognize the dangers of political failure and, on occasion, seek to protect policies from subversion.

The clearest example of the legislative process searching for some defense against political failure is the greatly debated Gramm-Rudman Act. This is a case where Congress desires to enact a national mandate--cut the federal deficits. A majority of representatives and senators understand, however, that we cannot simply rely on unconstrained politicians to act in voluntary concert to attain this goal. What we can do is take the goal of budget cutting and isolate it from the debate on where, in particular, the budget should be reduced. If deficits are not reduced enough when we tote up the thousands of items that Congress must address, no matter. Gramm-Rudman takes effect and the ultimate goal of shrinking the budget is preserved.

2.2. States of the World

In this section, we broadly review the current state of the world as it pertains to agricultural policy. Generally speaking, most observers have shifted focus from the income disadvantages of farmers to problems associated with instability. The structural bases of past policies are much less applicable in the 1980s. In general, farmers today are not economically "disadvantaged," nor are they in need of wealth transfers from the rest of society. It is widely recognized, however, that the links between agriculture and macro-economic and international conditions have frequently exposed farmers to intolerable fluctuations in financial costs, land values, and returns to farming.

There is a general feeling that traditional agricultural programs have lost much of their relevance in the presence of the wider economic environment. Policymakers may view with at least mild despair their recent efforts to intervene for the benefit of the farm sector. They have designed an expensive but seemingly ineffective set of policies; and they have witnessed an increasing concern by a widening variety of groups--from commodity producers to input suppliers, banks, and conservationists--all in the context of growing pressure to reduce federal expenditures and move closer to a deregulated economy. Brief examinations of several relevant dimensions of the economic and political environment follow.

2.2.1. Foreign Policy

While particular aspects of our foreign policy may seem incoherent at times (especially when the country is searching for a consensus on certain issues), the broad principles of policy (the objectives and strategies) are generally constant and well known. After all, the problem of how a state conducts itself in the world has been around for a long time. In contrast, we do not have centuries of experience when it comes to defining what we want for agriculture and the path to achieve these desires. At the same time, agricultural policy is often overwhelmed by foreign policy concerns. An agricultural policy designed to succeed under one set of circumstances today may turn out to be strongly contradicted by how foreign policy must respond in the future.

The infamous Russian grain embargo is popularly repeated as an example of how the interests of agriculture are ignored when the country thinks about foreign policy. Whether or not agriculture suffered greatly by the embargo is a matter of debate, but it is a sharp lesson on where agricultural policy fits into the bigger picture of national interest. It also demonstrates how easily a policy—especially an inflexible one—can be thwarted by immediate changes in the government's foreign policy tactics. Agriculture would be wise to remain circumspect in pushing its interests with respect to foreign affairs. When President Carter viewed his possible responses to the Soviet Union's invasion of Afghanistan, he put little weight on the opinion of American farmers who could sit safely and criticize his decision.

Food aid to less-developed countries and export subsidies seem quite consistent with farm income supports but tend to conflict with consumer interests. Income supports, however, through loan rates and the like, tend to conflict with the long-term goal of free trade. Furthermore, the unrestricted flow of commodities across borders tends to increase the instability of agricultural prices and farm incomes. The European Economic Community (EEC) has a policy of inducing increased food production, which directly affects American farmers, by taking away world markets and lowering prices. What our government wishes to do about this is not just an agricultural issue but a question of how we respond to our friends when they do something that hurts our farmers. The United States wants to promote the development of other countries, which means helping their own agricultural economies, sometimes at the expense of American markets. How can we, for example, jibe the development of the Caribbean Basin countries with the restrictions on one of their main exports--sugar? It would seem an embarrassing assignment for a U.S. diplomat to explain to a country, facing a world sugar price several times lower than our own, why we need quotas to maintain a concentrated number of prosperous, domestic producers who are relatively less productive anyway.

2.2.2. Public Health

Health issues, those unrelated to environmental concerns, have only recently begun to conflict with agricultural policy; but we predict they will become increasingly important in the future. Public health policy has always been extremely important; and agricultural policy, as we know it today, was designed to reflect this. The government inspects and otherwise controls the quality of much of what we eat, and it regulates to some degree the kinds of chemicals going into the production of food.

The public's perception of what the breadth of health policy ought to be is changing, however. Should not the government discourage the consumption of eggs and red meats if the consensus is that these foods are unhealthy? Why not prohibit or make very costly the use of tobacco if the medical and social costs of this legal drug are so great? These and similar questions will be heard more often in the future, and many farmers are not going to like the answers.

On the other hand, it is interesting to note that many agricultural programs that, sometimes on the surface, seem harmful to consumers' interest may actually be contributing to the goals of health policy. The tobacco program restricts supply and raises cigarette prices, which discourages smoking. Quotas on sugar imports do something similar for sugar-coated breakfast cereals and soft drinks. Programs that raise the price of feed grains make red meats and eggs more expensive relative to a vegetarian diet. Cheap water policy in California has increased the availability and decreased the price of fresh fruits and vegetables. These are examples of unintended consequences; and they stem, we think, from the incoherence of policy rather than from forethought. In general, however, we cannot rely on serendipity.

2.2.3. Tax Policies

Tax policies affect economic decisions: They can both

encourage and discourage investment; they may promote some industries and make others unattractive. Furthermore, while the enhancement of social goals usually justifies the implementation of taxes, the enrichment of special interests often results. Many recognize that the current, complex system of taxes explains much of the state of agriculture.[2] The recent roller-coaster ride of land values and the incentives to invest in larger, more capital-intensive operations can be traced indirectly to elements of this tax structure.

Today's taxes simply provide an environment for investment that is not neutral across industrial sectors, nor even between individuals in farming. Artificially short periods of depreciation, capital-gains provision, investment tax credits, and farmers' use of cash accounting are four principal elements of the present system that entice investment into agriculture, less for productive purposes than for avoiding tax burdens elsewhere. Nonfarm investors, especially in times of inflation, end up competing for assets (most notably land) with traditional farmers. Naturally, when conditions change and agricultural assets lose their tax-related appeal, many farmers remain holding greatly depreciated investments. These elements also tend to benefit larger farms with surer incomes and better collateral, relative to smaller farms and those with fewer capital resources on which to draw.

When farm investment is made so attractive in these ways, the capital intensity of agriculture grows. The industry moves toward a concentration of productive assets in the control of fewer, bigger concerns. The present tax policies conflict severely with goals of preserving family farms, maintaining rural communities, and generally protecting our traditional rural landscape.

2.2.4. The Structural Characteristics of Agriculture

It is widely recognized that the structure of farming has changed con- siderably since the introduction of large-scale governmental intervention in agricultural markets. Today, there are slightly more than 2 million farms, a third of the number 50 years ago; and, although the rate of decline in total number has slowed, fewer farms are expected in the future. Coincidentally, the average farm size has increased in terms of acreage, sales, and value of assets.

Knowledge of current and past farm structure is fairly complete, and a great deal of time has been spent in examining the topic as it relates to problems that future agricultural policy might address.[3] Information is scarce, however, on the connection between policy choices today and tomorrow's structural characteristics.

The present distribution of farm size can be most usefully characterized as trifurcated. There are a large number of small farms (sales of less than $40,000) contributing a relatively small amount of total output value. These farms represent, on average, an insignificant source of income (in some instances negative) for their owners; and they are minimally affected by agricultural policies. In part, their large numbers can be attributed to the official, and somewhat misleading, manner in which farms are defined. Over half of what are called small farms have sales less than $5,000 and four-fifths have sales less than $20,000. The majority of these small farms cannot be considered commercial enterprises but, rather, hobby farms or supplementary sources of income for rural families.

By contrast, there are a small number of large farms with gross sales of $250,000 or more, making up less than 5 percent of the total number. These farms produce nearly half of total output in terms of value, and they gain disproportionately from government aid. More strikingly, farms of over $500,000 in sales produce approximately 30 percent of output value but make up 1 percent of the total number of farms. Some large farms may be in serious financial trouble today. A consensus exists, however, that--with or without traditional farm programs--the largest farms will continue to dominate in agricultural production, although not necessarily in exactly their present form.

Medium-size farms, the remainder of the total, represent approximately one-fourth of all farms. They have sales between $40,000 and $250,000 and contribute slightly less than half to the total value of production. These enterprises correspond most closely to the traditional concept of the family farm, being the major occupation and livelihood of their owners. Program benefits generally flow to those who produce the most, but it may be the viability of the midsize farm and the welfare of midsize farm families that are the most affected by agricultural policies.

One of the major questions for policymakers is: How do programs affect farm structure--especially farm size?[4] While there have been significant changes in structure, the implications of agricultural policy are not well known. Theoretical analysis seems to point to governmental programs inducing increases in farm size. Empirically, however, this is difficult to verify. Large farms do benefit the most monetarily, due to payments based on production levels. For instance, in 1985, farms with sales over $250,000 received 32 percent of direct government payments. In addition, families with specialized large farms are more dependent upon farm income as opposed to off-farm income. Therefore, programs reducing agriculturally related risk (programs dampening price fluctuations and insuring yields) would tend to make such farms more attractive enterprises. On the other hand, the cost advantages associated with technology of large farms may be the primary determinant of long-term profitability relative to smaller sizes. Hence, farm programs available to all size farms have merely added a margin of profit--the greatest additional profits going to the biggest producers.

In terms of income and assets, farming as a sector is not as badly off as popular conviction sometimes holds. There are individual hardship cases plagued not so much by income problems as by financial stress. Compared to the past, when farm policy was motivated by widespread hardship and dislocation, income is currently not as important a topic as it once was. Aggregate farm income is lower than previous levels, but this is not a good representation of the welfare of individual farmers. Average farm-family income is approximately $30,000. Off-farm income makes up about 60 percent of this average level. But even average income is a poor indicator of family welfare. For farms with less than $40,000 in gross sales, net farm income is usually a minor component of average income--in fact, averaging a loss of $1,635 in 1985. In the same year, farms between $100,000 and $500,000 in sales had an average income of about $60,000 with approximately one-sixth of income from off-farm sources.

The current financial stress suffered by some farmers has brought greater attention to the debt-asset position of farming as a whole.[5] The agricultural sector has over a trillion dollars' worth of assets, primarily in land. Land values, however, have declined dramatically, bringing asset values down. As land prices increased during the past decade, debts also swelled. Today, the debt, relative to net farm income, is almost twice as high as 15 years ago.

Debt, relative to assets, increases with farm size;

and large farms have, on average, the least enviable debt-asset position. Even so, the debt/asset ratios are more favorable than for comparable nonfarm firms. Moreover, large farms have both greater incomes and asset values, which give them some resilience in handling financial hardships. Families with small farms also are somewhat insulated from financing problems because they have a greater proportion of incomes from off-farm sources; and they, for the most part, avoided the purchase of land at inflated prices in the 1970s for the purpose of expanding production or speculating on further price increases. It is the medium-size farm that typically is at the mercy of future income instability and asset devaluation.

While the income and financial problems of some farmers are of much immediate concern, the issue of productivity is one of general and long-range interest. Productivity has been given considerable attention recently since it will influence future supply and the ability of U.S. agriculture to meet domestic and international food demand.[6] Greater cost efficiency will tend to benefit farmers but at the expense of depressing prices and perhaps adding to what is thus sometimes called excessive productive capacity.

Agricultural productivity has increased substantially in the past 50 years due to public and private investment in research and to increases in farmers' managerial ability. The taxpayer has heavily supported productivity research and has reaped significant benefits in terms of lower commodity prices. The growth rate in farm productivity is projected to decline in the future, but this is not a certainty. More efficient methods of pest management and changes in input use (water and fertilizer) and tillage may contribute to increases in the productivity growth rate. The greatest unknown as yet--and one of the more important gaps in policy analysts' information--is the effects brought about by biological-engineering technologies.[7] Some specific farm products (for instance, milk) may have further increases in productivity that far outpace agriculture as a whole. The resulting boom in the yields of particular commodities may drastically depress prices and place severe pressure on governmental programs designed to regulate supply for the purposes of enhancing incomes. And, as in the case of dairy uses of grains, productivity changes in one agricultural industry may have detrimental effects on another. Consequently, a wide range of governmental programs may be affected ultimately.

2.2.5. Macroeconomic and International Links

The performance of the agricultural sector is determined, in part, by the larger economic system comprising the macroeconomy and international economy. This has been—at least qualitatively—understood for some time. The importance, however, of these macroeconomic variables has not been fully appreciated or accounted for in the construction of traditional agricultural policy, at least until recently.[8]

In addition to the divergence of professional perspectives, economists and policymakers lack concrete information by which the importance of macroeconomic links can be judged or tested. For instance, only recently have interest rates exhibited significant volatility, allowing observation of consequent responses in the agricultural sector. It is disheartening to note that the information necessary for policy choices is being revealed by the very conditions causing such concern.

Despite the spareness of hard empirical analysis, we know that there has been a greater interdependence between national economies (in terms of both volume of trade and capital markets) and a significant change in the farm sector's relation to credit markets. Agricultural exports make up approximately one-fifth of the total value of total U.S. exports. Furthermore, net farm-product exports are consistently positive compared to a net deficit for non-agricultural goods. As the agricultural sector has grown more dependent on exports, the nature of aggregate foreign demand has become an increasingly important question. One of the major issues not yet fully understood is whether export demand is sufficiently elastic for a decrease in prices to be accompanied by sufficient increases in volumes, thereby raising long-run total income.

We also know that the agricultural system is extremely sensitive to interest, inflation, and exchange rates. The agricultural sector is more than twice as capitalized as manufacturing on a per worker basis, taking account only of physical capital, not land. A broad consensus holds that current governmental deficits, and the reluctance to monetize the debt, have maintained real interest rates at levels debilitating to many farmers. There are two main aspects to this issue. First, the current debt-asset position of some farmers leaves them financially strapped—exposed to intolerable cash-flow and equity problems. This is a relatively direct aspect of agriculture's links to the larger economy through interest rates. Traditional commodity programs are ineffective at addressing this condition. Second, less directly, high interest rates induce a strong dollar, which reduces foreign demand for U.S. exports and increases import competition. A weak dollar in the past aided exports by keeping prices to foreign demanders low. Commodity programs, tending to support prices, exacerbate any reduction in export demand and fail to pressure a necessary contraction in U.S. production.

The effects of federal fiscal and monetary policies on interest rates, exchange rates, and inflation must be placed in the context of other international trade issues. The trend in world agricultural trade is toward a greater dependence among nations, greater competition between suppliers, and lower export prices generally. The world recession and associated international credit problems brought about a reduction of demand for U.S. exports, exacerbating the effects of high domestic support prices and exchange rates. In the minds of many, the continuing world recession will be the most important obstacle to revived growth in world agricultural markets. In addition to world economic conditions, an indirect effect of high interest rates and a strong dollar may be to encourage foreign governments to contract their own money supplies. This, in turn, leads to lower aggregate foreign income and a lower demand for U.S. farm goods, at least in the short run.

Future levels of foreign demand are likely to grow at a slower rate than the United States experienced in the 1970s, although faster compared to the recent past. Slower growth in export demand is attributable to several conditions. First, world population and economic growth rates are likely to decrease. Second, an increase in foreign production will be encouraged as a result of continuing development, technological improvement, and the improved exchange rate and trade positions of other producing countries. Increased agricultural production in some nations may not be negative per se for U.S. producers when viewed from a longer term perspective. Indeed, foreign agricultural development offers some hope for U.S. exports. Farm sector development may lead to increased incomes and increased demand for certain U.S. commodities. Whether or not, in general, foreign agricultural development bodes well for U.S. exports has not been determined.

A third condition dampening U.S. export demand is the maintenance of uncompetitive agricultural and foreign trade policies in many countries. Several nations restrict potential imports from the United States by use of a number of barriers; they may even be behaving strategically to receive a price lower than they would otherwise. Exporting countries will likely continue to market their farm products more aggressively than the United States.

In the early 1980s, four factors of the domestic and international economy came together to produce severe pressures for an adjustment in U.S. agriculture. High interest rates, a strong dollar, a contraction in world income and demand, and institutional barriers to trade--all indicate that resources should move out of U.S. farming to get the sector into equilibrium with the rest of the economy. However, because of agriculture's capital intensity and its major dependence on international trade, this combination of factors has meant that farmers have had to pay a painful adjustment tax. This tax not only took the form of higher interest payments and lower commodity prices for goods whose supply was not shrinking fast enough but also reduced farmers' stock of wealth--wealth that had accumulated in property values over a period when macroeconomic and international economic conditions were more favorable.

A final major issue is whether the international market is becoming more or less stable. Greater stability seems to be implied by several factors: the increased use of forward contracting markets to anticipate prices, the more predictable behavior of major importers (e.g., the USSR), and the greater integration of national markets. The recent history of agricultural trade, however, is not encouraging. Instability in the future may result from increased production variability in the United States and abroad and increased uncertainty about domestic and foreign policies.

To be sure, regardless of whether international markets are more or less unstable, U.S. markets can be more unstable because of large shocks to financial markets, exchange rates, and international commodity markets.

2.2.6. Resource and Environmental Dimensions

Traditionally, in discussions of agricultural policy, resource and environmental issues have been given much less attention than farm structure or macroeconomic and international links. In part, this is due to the long-run nature of resource problems--there is a lack of immediate concern. Nevertheless, interest in soil erosion and water quality especially has grown and will continue to grow as better information on such problems becomes available an the focus on farm welfare shifts from short-term income difficulties to long-term productivity maintenance. Many analysts and policymakers are starting to anticipate a need to integrate agricultural and resource policies.[10] Recently, a major concern has been expressed for land degradation--erosion, increased salinity, and conversion of cropland to nonagricultural uses. The general consensus is that soil erosion has increased for the worse, particularly during the 1970s as crop acreages increased. The on-farm costs of erosion have overshadowed off-farm costs (e.g., sedimentation), which are perhaps much greater. Knowledge, however, regarding these costs and how to treat them is limited, though increasing. Data on off-farm costs are necessarily harder to obtain, and there is some confusion over how best to treat off-farm problems related to erosion. While reduction of farmland erosion will in some ways reduce off-farm costs, it is possible that direct treatment of off-farm problems may be more socially profitable. Additionally, practices aimed at controlling erosion may actually exacerbate water quality problems due to increased runoff of farm-related chemicals.

Excessive land degradation occurs on a small proportion of total agricultural land (less than a third), but these losses are highly concentrated geographically.[11] The general view appears to be that government policies are warranted by the divergence of social and individual interests. Past policies may have encouraged farms to maintain production of supported crops on erosive land--crops (feed grains, soybeans, and cotton) that take up most of the land with erosion problems. In addition, the government has provided at least implicit import subsidies in the form of natural gas, water, transportation, and other facilities; these factors, too, have resulted in the expansion of farming to the detriment of resource conservation. There is also a macroeconomic link to resource use: Higher interest rates today signal farmers to place a greater value on current proceeds from additional output. Simultaneously, a reduced interest in resource-conserving investment is created. Future increases in land productivity may induce reduction in acreage and thus in soil erosion. Previous technological changes have emphasized the expansion of crop acreage. Much improvement, without greater public expenditure, may be had, however, by simply increasing information to government agencies and farmers and thereby improving land management and erosion-damage control.

The conservation of water resources may be of even greater immediate concern than that of land. The agricultural sector is, by far, the thirstiest consumer of water in the United States, accounting for over 80 percent of total use. Surface and groundwater irrigation is applied to 25 percent of total farm production and is especially important in western regions where most irrigated acreage lies. A primary problem is the nonrenewable nature of some water supplies.[12] Increasing irrigation and declining water tables have resulted in increasing costs to farmers. Coupled with the quantitative aspect, increasing amounts of farm-related chemicals and wastes have contributed to agriculture's burden on water quality, especially the rising salinity levels.

In addition, future increases in nonfarm water consumption will undoubtedly bring greater pressure to change the system that once offered cheap and abundant supplies. Farmers and public agencies will need to develop less water-intensive systems as the resource base tightens. Present and future policy must anticipate these transformations in the state of natural resources and recognize the changes in prices and other incentives that will be required to effectively manage this common property resource.

2.2.7. Specific Commodity Settings[13]

Few government programs that invite contention are broadly aimed at agriculture in general. Most are designed to aid particular commodity groups and, therefore, affect differently a variety of economic and political interests. The emphasis on individual commodity programs has contributed to the growth of numerous interest groups that take part in the policymaking process. Food and feed grains (and, in part, cotton and rice) are traditionally the main considerations in policy debates due to their widespread production, volume and value of output, and the large number of farmers involved. These products are also regarded as "basics," giving them an aura of historical importance for society's welfare. In addition, food and feed grains make up a significant proportion of exports that have recently occupied—and will continue to occupy—the attention of agricultural policymakers. Macroeconomic and international links are likely to be principal concerns for producers of these commodities. More than for most other farm policies, the integration of nonagricultural variables (interest and exchange rates) into program operations will be considered in the future.

Food and feed grains are the biggest recipients of Treasury transfers to the farm sector. Acreage controls and price supports have been used extensively for many years, although their effectiveness has been increasingly questioned. For the immediate future, grain prices are unlikely to rise significantly, and demand will remain weak relative to the past decade.

Dairy policy has lately acquired some degree of notoriety. The general feeling in the community of farm policy watchers is that "something will have to give" in future dairy programs. Until recently, production has continued to grow faster than consumption, and U.S. Department of Agriculture purchases have been high. The dairy industry is characterized by marketing orders and price supports which have been motivated by desires to stabilize prices and enhance incomes. Yet, price supports have risen faster than the general price level, and marketing orders are perceived to effect collusion. The very success of dairy producers in influencing government programs for their benefit in the past will make them highly visible targets for policy changes.

Past government programs to control supplies have been almost certainly capitalized into the value of the resources used in milk production, especially dairy cows. The public has borne the costs through higher prices and storage expenses. Recently, however, payments have been made to reduce the level of production, thus lowering the high cost of restricting supplies to consumers. Traditional supply manipulation will become even more difficult and costly as technological advances contribute to higher yields.

Sugar is another commodity with a somewhat infamous popular reputation. It is also unusual, although not unique, because government support takes the form of import quotas that increase domestic prices. The present U.S. price is several times that on the world market, revealing plainly to the public the cost of intervention. Complicating the issue is the increasing profitability of corn sweetener substitutes for commercial sugar. This tendency has been noted, but the links between corn and sugar policies have not been well examined.

Sugar is produced in this country by a relatively

small number of beet and cane farmers. The concentration of benefits to a few producers, despite the significant but diffuse costs, has often been used to explain the continuation of policies that create considerable long-term allocational inefficiencies and losses in consumer welfare.

To be sure, there are rationales given by sugar producers for sugar programs as they currently exist; but the pointed discrepancy between domestic and world prices and the large costs of programs will make sugar, like milk, a commodity of particular scrutiny in the future. World production is likely to grow and any continuing strength of the dollar relative to other currencies will create pressure for increased imports and lower domestic prices.

The livestock sector has suffered for some years in the environment of unstable grain prices and uncertain public policy. The sector is comparatively free of direct governmental intervention, although commodity programs have a significant effect through the influence of feed grain supplies and prices. Import restrictions are in place, aiding livestock producers by reducing competition and raising prices. These restrictions, however, primarily affect lower quality, lower priced meat products. Thus, low-income consumers are harmed disproportionately.

Commodity programs have two effects on the livestock sector. First, programs tend to raise average prices of grains by offering incentives to reduce production or by restricting supplies available to grain consumers. Second, programs influence the stability of grain prices facing livestock producers by dampening the variability of grain supply flows. This stabilizing effect of commodity programs must be balanced against the destabilizing role played by frequent changes in farm policy. For example, the introduction of the farmer-owned reserve led, in part, to feed grain price increases unanticipated by the livestock industry. The effects on industry dynamics and the costs of adjustment have gone largely unnoticed.

The livestock sector in general would undoubtedly like to see stable, low prices for feed grains and protein; but it would benefit from policies that trade off price stabilization with higher average prices. The recent past, however, has been, if anything, more uncertain and disadvantageous to the industry because of policy uncertainty. Future policy must also take account of the shift in consumption from red meats to chicken and fish.

This trend may be due not only to changes in tastes and health concerns but also to recent volatile red meat prices and squeezes on consumers' discretionary income.

In terms of governmental policy, tobacco is a unique commodity due to the use of quota allotments and the manner in which program costs are sustained by producers themselves. Historically, Treasury costs have been kept low as a result of movable quotas tied to stock levels. The tobacco industry is characterized by small-farm acreages and a traditionally high labor intensity. Elimination of the present policy probably would shift the structure of the industry to fewer and larger cost-efficient farms. In addition, a free market for tobacco would remove the benefit now going to owners of quota allotments who typically are not the producers of tobacco. In the market environment of quotas, the supported price of tobacco and high exchange rate of the dollar have led to increased imports; a change in policy that permits increased production would probably reverse this trend and make U.S. tobacco more competitive in the world market.

2.2.8. Other Groups' Interest

There are several nonfarm groups that have a particular interest in agricultural policy.[14] Input suppliers benefit from increases in production spurred by increased demand or government inducements. They suffer, as has happened recently, from reduced production. Government programs to slow farm output growth are understandably disliked by producers of seed, fertilizers, machinery, and the like. Their recent experience with the payment-in-kind program (PIK) has revealed the potential hazards of being left out of the policymaking process.[15] Input suppliers have two goals in mind: increase output to increase the demand for their goods and promote high farm income to allow farmers to invest and respond to higher commodity prices by increasing supply. Farm policies designed to raise incomes by reducing acreage or otherwise restricting output, therefore, will be opposed by input suppliers who would rather see output-increasing and alternative kinds of income-enhancing policies.

Banks and credit institutions are, in particular, adversely affected by farmers' financial and income problems. Stable farm incomes are more important to this group than merely high average incomes. The recent depreciation of farmland has caused severe financial stress to some farmers and reduced the overall incentive for farmers to make use of credit opportunities. Creditors desire agricultural programs that stabilize farm incomes and protect against future financial crises.

Consumers, on average, are better off now than in the past. Commodity prices are relatively low, and consumers have an interest in keeping them so. Programs that restrict supply and raise prices to increase farm incomes are harmful to consumers, although the diffusion of costs over many consumers usually makes opposition to such programs ineffective. Like input suppliers, consumers would prefer policies that avoid supply controls; but price stability is also desired, making governmental intervention in supply flows more attractive. In general, however, consumers as taxpayers consider income transfers to the farm sector detrimental. The sensational budgetary costs of PIK and the current problems of financing the federal government have reinforced the public's doubts about farm policy.

Consumers also have an interest in continuing productivity growth—both in the United States and abroad. Federal support for productivity research has protected consumers from rising food prices. Indeed, the trend has been toward ever-decreasing real food prices as agriculture has grown more productive. In addition, consumer and farmer interest may coincide in some policy efforts to expand demand. Food stamps, for example, have been treated both as a partial solution for farm income enhancement and as a benefit transferred to low-income consumers. In broad terms, however, consumers' interests are more apt to coincide with those of assemblers, processors, distributors, and wholesalers. These intermediaries operate, for the most part, with small margins and large volumes; and they would benefit from policies tending to increase production and lower prices.

2.2.9. Political Dimensions

There has been a general trend in recent years toward deregulation of the economy. This has originated in a growing respect for the efficiency of the marketplace and a disenchantment with the government's ability to improve on that efficiency. The agricultural sector has been protected from the trend toward deregulation, but it has not been immune.

Despite the attractions of deregulation, there is another development that complicates policymaking. Over many years, a participatory (or more pluralistic) democracy has emerged. Interest groups now find it much more profitable to engage in efforts to influence the working of government. No longer do representatives of the people make laws reflecting the relative strengths of their constituent voters. Elected officials and bureaucrats now are lobbied by well-financed groups to effect changes in, or prevent alteration of, the complicated machinery of the state. This is still done in the confines of voter approval, but political power more often reflects narrow interests in Washington than the desires of regionally dispersed voters.

The continued growth of interest groups in agricultural policymaking represents an increased likelihood that programs designed to attain broad social goals will be altered to reflect the relative lobbying power of narrowly focused groups. Discussion of agricultural policy should thus take the possibility of political failure into account, or programs could be designed that contain the seeds of their own failure. An example: The farmer-owned grain reserve program was originally planned to stabilize commodity prices and supply; but, following the Soviet grain embargo, it was used to remunerate farmers by absorbing supply and thus increasing prices to raise incomes. Hence, a program for stabilization became a program for income transfer to the agricultural sector. The result was an ever-growing amount of stored commodities that hovered over the market and placed an intolerable cost burden on the government. The PIK was born as an emergency measure that shifted the immediate burden of program cost from taxpayers and farmers to others--input suppliers, especially.

2.3. Frequently Debated Policy Alternatives

In this section we review a number of broadly defined alternatives that frequently arise in the general debate over agricultural policy.[16] Table 1 (Tables at the end of chapter) presents a synopsis of the intended objectives of several proposed policies, including our own (detailed in the next section). A marked box indicates that a policy is motivated by the corresponding goal. This, of course, does not evaluate the realized effects of implemented programs. To be sure, programs would influence the entire range of objectives and their influence would depend upon the economic and political environment. The broad alternatives we discuss in the following several pages are free market, reinforced free markets, revenue insurance, flexible loan programs, supply contraction, and demand expansion.

2.3.1. Free Markets

Historically, the inability of an unfettered market to attain the social objectives outlined in section 2.1 has been the principal justification for federal intervention in agriculture. Dissatisfaction with governmental policies and programs, however, has called into question whether tampering with the market leads to problems worse than those that motivated public action in the first place; as political attitudes and goals have changed, there has been a growing respect for free-market outcomes. Unsuccessful governmental intervention may be traced either to the technical infeasibility of a public policy or to an adulterated translation of social goals into political reality--the result of political failure. Whether for good or bad, future policy proposals will be more critically judged, relative to the expected results of unregulated private enterprise.[17]

Deregulating agriculture would shift the distribution of wealth from one determined, in part, by the ability of farmers to take advantage of federal programs to one determined primarily by the ability to manage and produce efficiently in a free-market environment. Inefficient producers certainly would be pressured to leave the sector; their wealth, resulting from the capitalization of government-sponsored rents in land prices, would deteriorate. Efficient managers would reap most of the benefits of a policy of inaction, and taxpayers would be relieved from supporting prices. If farming exhibited increasing returns to scale, the tendency to large operations would be accelerated. As mentioned above, however, there is evidence of increasing returns to scale in the production of government benefits. The structure of agriculture may, therefore, move to smaller units under deregulation.

The most drastic effect of a free market may be on the variability of farm incomes. Present programs do isolate agriculture from severe changes in the economy and weather. Uncertainty in agriculture, however, is not due only to weather and the working of the marketplace; there is also the uncertainty associated with the political system of farm programs. There is a trade-off between market uncertainty and policy uncertainty.

Whether or not private storage would compensate for the reduction or loss of government-sponsored storage is an empirical question, but the absence of deficiency payments and loan rates would almost certainly create greater instability in supply. This point also relates to the objective of food security. If private storage does not completely replace government storage, a policy of free enterprise may decrease the provision of adequate safe food reserves.

Moving to a free market for agriculture would increase the sector's capability to respond to long-run changes in economic conditions. Changes in market conditions--in this country as well as the rest of the world--would be quickly reflected in the returns to farming. Therefore, a more flexible and efficient agricultural sector would be encouraged, avoiding occasional policy crises associated with governmental attempts to shield farming from world market conditions. In addition, the ability of U.S. agriculture to compete efficiently in world markets would be enhanced by the elimination of policies that maintain artificially high prices.

As noted above, several past policies that attempted to avoid the outcomes associated with free enterprise have been subject to policy crises. Reliance on free markets would avoid such costs; successful producers would be determined by their productive, not political, efficiency. In addition to avoiding political failure of implementation, a large burden on federal expenditures would be eliminated.

2.3.2. Reinforced Free Markets

The disadvantages of ending federal involvement in agriculture have led to several proposals designed to overcome the deficiencies of a free market. The major problem of complete deregulation is the increased volatility of prices and farm incomes that would result and the possible displacement of a sizable number of farm operators. Proposed policies addressing this issue take two forms: governmental sponsorship of risk-trading institutions and direct governmental involvement in dampening shocks through commodity storage.[18] Both forms of policy would maintain the essential advantages of nonintervention while reducing the degree of uncertainty facing producers.

Personalized insurance against adverse crop yields is now available. Insurance against adverse price movements has been provided by government loan-for-storage and deficiency-payments programs. These programs have not only set a floor on prices but have been used as income supports as well, isolating agriculture from the realities of the marketplace. Moreover, past programs have not been designed for individual farmers nor have they attempted to charge producers for society's cost of absorbing risk.

Futures markets do afford farmers hedging opportunities. Nevertheless, these markets have proven unpopular with farmers due to the short term of existing contracts, imperfect capital markets, and the degree of exposure to margin calls. The use of futures contracts can be encouraged by lengthening contracts to give price protection for two to three years. (The market, however, would be extremely thin for long-term contracts.) In addition, the government could intervene directly in futures markets, reducing the variability of contract prices and thus decreasing the exposure of everyone in the market. Put options have been suggested as a better vehicle for farmers wishing to hedge against price decreases. (A put option is the right--though not the obligation--to sell a good at some specified price in the future.) Purchasing options would remove a farmer's exposure to a long string of margin calls. In addition, the cost of options could be subsidized to encourage their use.

Any attempt to reinforce the workings of a free market is necessarily subject to political manipulation. For example, once in place, a system subsidizing the use of put options for farmers could entail a net wealth transfer to agriculture. In some sense, overly subsidized insurance would induce overly risky production behavior by farmers. Establishing stock subsidies for price stability and food security also leaves open their later political manipulation as a means of enriching storers.

2.3.3. Revenue Insurance

Revenue insurance is an appealing idea in theory and has attracted much attention, but proposals for this plan are still lacking in specifics (Congressional Budget Office, August, 1983). A farmer would choose to insure gross revenues at some level, the premiums being based on this level and on farmer and farm characteristics. Revenue insurance would be a more individualized risk-management tool than the present system of programs. Looking only at private benefits and costs, a farmer would select a desired level of insurance rather than accepting a package of loan rates, diversion requirements, payment limitations, and other program restrictions.

Farm revenue is difficult to insure: If the price is low for one farmer, it is low for all farmers. This exposes an insurance company to great financial risk and makes large premiums imperative. Moral hazard further complicates the insurance problem: Once a farmer insures his revenue, there is little incentive to allocate resources appropriately. For example, to avoid some of the problems of moral hazard, premiums could be made contingent on the continuation of historical average yields. Revenue insurance would also have to contend with adverse selection. Insurance would be most attractive to those who are least adept at risk management; and those experienced in other means of risk management, such as the futures market, might not purchase revenue insurance. Despite these difficulties, private insurance companies could offer revenue insurance; but high premiums would severely limit farmer participation.

2.3.4. Flexible Loan-Rate Policies

Commodity storage programs are intended to moderate price fluctuations by the accumulation of government stocks or by the subsidization of private storers. Traditional stabilization tools, however, have been notoriously unresponsive to market signals, burdening farm policy with costly and sometimes embarrassing levels of stocks. In response to this problem, the recently implemented system of storage programs are flexible (i.e., market-conditioned) loans. Loan rates are responsive to market signals and vary, depending on the economic environment.[19]

When program instruments are left unadjusted as the economic environment changes, a policy disequilibrium develops: the tools are no longer appropriate for, and perhaps contrary to, the original policy objectives. This leads to a policy crisis when dissatisfaction with either program benefits or costs is so widespread that a change in policy becomes inevitable. Under the 1985 Food Security Act, flexible loan rates, however, change as the economy changes, reducing the likelihood of policy crises.

2.3.5. Supply Contraction

Past governmental intervention in commodity markets has tended to support excess supplies through the maintenance of above-market prices. Price supports do serve the purpose of raising and stabilizing farmers' incomes, but the government is left with the cost of storing large amounts of grain. The ability to manage ever increasing stocks is limited, both financially and politically; the inevitable result is a change in policies and an increase in the uncertainty regarding governmental action. Recognition of this tendency to increase commodity supply has led to acreage controls, marketing orders, and pressure to divorce income support from production. Acreage controls are only crudely effective because supply does not correspond exactly to the amount of farmland under production--the so-called slippage problem. Marketing orders traditionally have been applied to agricultural commodities where producers are regionally concentrated and easily organized. Orders also tend to weaken farmer independence and are popularly perceived to effect collusion, as in the case of milk.

Different methods of supply contraction, both mandatory and voluntary, have been used since 1933 in efforts to increase farm returns. Restrictions of supply can be used alone to raise market prices or to counter the incentives to increase output provided by production-based, income-support policies. The voluntary acreage set-aside program is the current policy tool; but due to free-rider and slippage problems, it does not effectively reduce production. Given the limited results, it has proved to be a rather expensive program.

2.3.6. Demand Expansion

Demand expansion is popularly perceived to be an easy solution to the problem of low farm incomes brought about by low prices. Other historically popular policy alternatives for raising incomes are supply restriction, which is unpalatable to farmers who must cut back production, and direct payments to farmers, which are costly to the government and too obvious to be politically attractive. Demand expansion, however, does not typically require any unpleasant adjustments by U.S. farmers. Unfortunately, private domestic demand and government demand are unlikely to expand significantly (particularly after PIK), and exports of major commodities decreased substantially in the early 1980s. Most proposals for expanding demand concentrate on the foreign market. This is, however, not a perfect solution; as farmers expand into foreign markets, they face more price instability because prices are subject to international shocks.

One long-range policy to expand export demand is to use U. S. agriculture's strength and abundance to improve economic conditions in less-developed countries. Many charges have been made that subsidized exports to poorer countries (P. L. 480 type programs) in fact undermine production in recipient nations. Nevertheless, special, carefully crafted aid to poorer countries in the form of agricultural products or long-term credits can foster economic development.[20] This would offer immediate outlets for U. S. commodities while improving U. S. export prospect in the future.

In addition, there are two more direct approaches to solving the problem of low foreign demand for U. S. commodities: either lower the domestic price of commodities or maintain the high domestic price but subsidize the export price directly through credits. Lowering the domestic price is not, in general, a popular policy; therefore, subsidized export plans have generated much interest.

3. A Coherent Agricultural Policy

The previous sections have illustrated the incoherence of current agricultural policy, the goals that policy should pursue, and the environments (or states of the world) in which policy must operate. On this basis, we present our proposals for a more coherent and effective farm and food policy.

Our proposals are necessarily general. Instead of offering precise and detailed mechanisms, we focus on the basic motivating ideas and the ultimate implications of our policy design. Some of what we offer may be found in existing programs, some are variations on familiar themes, and some are novel concepts that should be publicly debated.

Our purpose is to design a prescriptive, coherent policy. We have given no thought whatsoever to political feasibility. Nevertheless, the possibility of political failure must be explicitly addressed if a new set of programs is to meet chosen goals in a coherent manner. The policy must also be insulated from bureaucratic discretion over programs and narrow legislative concerns.

Once the social debate over goals and policy is closed, there should be sufficient disincentive to change bits and pieces of policy for the purposes of undermining the large consensus. This is not to say that legislative discretion is undesirable or that further debate over basic issues should be avoided as mistakes and/or new signals emerge. Rather, if the U. S. Department of Agriculture or Congress wishes to alter programs, then the alterations must complement the original design. As is painfully evident in present policy, the incentives for small coalitions to take advantage of the potential for political failure have led to a badly fashioned policy.

Different parts may please different tastes, but the policy in its entirety pleases very few.

3.1. Specific Proposals

We turn now to specific proposals for a more coherent agricultural policy. The proposals are offered to address both the broad social goals outlined in the second section and the problems of political failure. Each program is designed to be widely applied, or available, to agriculture and closely connected to a specific objective.

Our newly proposed programs imply the elimination of commodity-specific programs as the basis of policy. Commodity programs have become vehicles for income transfers to larger farm operations, and they offer primarily a band-aid approach to farm problems. They are simply too easily manipulated by specific interest groups to be successful.

Commodity programs are the worst examples of the incoherence of present policy. Recently, for instance, we saw the Dairy Buyout Program attempting to deal with milk overproduction induced by governmentally supported prices; at the same time, the livestock industry was in court trying to stop the program in order to prevent a feared decline in meat prices due to the slaughtering of dairy herds. The conflicts generated by the present system of programs are not just between commodity groups. Within commodity groups, some farmers, often those who need little help, gain more than others. Target prices, loan rates, and diversion requirements can be influenced so that particular farmers may take greater advantage of possible benefits, while taxpayers and other farmers suffer the burden of "captured" programs. In essence, we propose that commodity programs be replaced by a set of programs available to all agricultural producers.

3.2. Minor Changes in Current Programs

Not all present programs are without value in terms of the objectives presented in section 2. In fact, several actually work and are worth maintaining, perhaps with minor changes. If these programs do conflict with other components of agricultural policy, then we should examine, change, or eliminate the other components.

Low-income food subsidies, for example, should be maintained. They go far in achieving social goals of equity, and they also expand the demand for food production. Certainly, there are anecdotal stories of defects in the Food Stamp Program, but there is sufficient evidence that it is effective. Society may consider revising the existing system of low-income food subsidies, perhaps even eliminating it in favor of more direct income supports. While changes in the present system might reduce food demand, it is less a matter of agricultural policy than basic decisions regarding social welfare.

We also propose that the system of food inspection and regulation be maintained largely in its current form. The prominence of public health goals not only reflect growing social concerns but tend to support consumer demand by maintaining confidence in U.S. agricultural production. Perhaps we should channel more resources into the regulation of those chemicals that may find themselves in products for which they were never intended. The incidents with heptachlor-contaminated milk in Hawaii several years ago, and on the mainland recently, are instructive lessons on the dangers of these inadvertent contaminations.

One set of government supports that we propose to maintain, but with altered emphasis, is that for agricultural research. We suggest that a greater emphasis be placed on the principle that the burden of research expense be proportionately shared, based on the benefits accruing to various groups. Taxpayer support for research is presumably based on the notion that benefits are widely spread across consumers and producers. Unfortunately, this is not always the case. In addition, we suggest that the government reduce its support of research that may potentially result in larger, more capital-intensive farms. The structural effects of some applied research is often contrary to the goals of maintaining the family farm, rural communities, and the aesthetic rural landscape.

In order to make these changes effective, we propose that potential recipients of large federally applied research grants detail, for the public's inspection, the economic consequences of whatever results from their research. We suggest a sharper review of proposed research, independent of researchers' technical competence, that estimates the socioeconomic changes that may ensue. This means an account of the winners and losers and the ultimate effect on the industrial structure of farming. Something like this is already in place where researchers underscore all possible benefits of their work. The trouble with the present system is the lack of purposeful integration of research with broader agricultural policy.

3.3. Newly Designed Programs

We offer six newly designed programs as the basis of farm policy that would replace the present system based on current commodity-specific schemes:

(1) Tax Policy. Structuring a new tax policy for agriculture is perhaps the most difficult challenge society faces in managing the farm system. Taxes are, by their nature, wide ranging with myriad effects both direct and indirect. In addition, tax policies could never be restricted to certain industries--designing a new one cannot be an isolated exercise just for agriculture. Anyone whose attention is focused on farming must remain circumspect in approaching a new tax system that "solves" agriculture's problems. We do, however, have a few broad proposals.

A better tax policy would encourage greater neutrality across industrial sectors in the economy. That is, it would treat food and fiber producers and nonagricultural businesses equally. Such a policy is needed primarily to discourage two tendencies. First, investment in agricultural assets often has a component unrelated to current production, especially investment in farmland. Investment of this nature separates the value of land from its current productive worth.

Second, the current tax system, in some ways, subsidizes investment in capital-intensive production. In particular, large operations can take greater advantage of these subsidies than smaller ones, further skewing farm structure from social goals. The bias toward capital gains, artificially short depreciation periods, and investment tax credits should be avoided.

(2) Targeted Income-Deficiency Payments. We propose that the government provide direct income transfers to farmers based on certain characteristics society wishes to maintain. A program of targeted income-deficiency payments would serve not only to reduce the instability of farm income but also to support the goals of equity and preservation of the family farm and rural community. Deficiency payments would be based on the value of farm-production assets controlled by the recipient. Such assets include farm machinery, farm buildings, and arable land owned and rented. It would exclude the farmer's house and other assets not directly related to farming.

The government would designate farm income, say, for a family of four, that would be maintained if the recipient falls within some range of farm-asset values. For example, the designated income could be $10,000 and the range of asset values for subsidization could be between $75,000 and $400,000. A lower limit for asset values is needed to exclude many who might otherwise be classified as farmers but enjoy off-farm sources of income. By making up shortfalls in farm income in this manner, society discourages farmers from moving the size of their operations beyond the limits eligible for the program.

The government could make use of farm appraisers to certify the value of productive assets. A farmer might reduce the "program" value of his assets by giving up ownership of land to family members and relatives, still retaining control over the land's productive capabilities. To prevent such arrangements, we would establish a rule that prohibits a farmer in the program from transferring land (by sale, lease, or gift) to family members unless the receiver of the land establishes independent control over the land. Severe penalties should be imposed on persons who collude in an attempt to satisfy the means test under the program.

(3) Anticyclical Credit Program. In order to reduce the instability generated by macroeconomic and international exchange rate fluctuations, we propose a government-supported credit policy for farmers that dampens the swings in agricultural investment. This particular policy instrument must be designed to address failures associated with incomplete risk markets, lack of equity capital, and "overshooting" resulting from sticky nonagricultural prices and the short-run nonneutrality of money.

Farm-credit system interest rates should move by an established rule in the same direction as agricultural prices. As prices rise, interest rates should also, dampening the expansionary effects on investment and land values. As prices decline, interest rates should fall, mitigating the farm credit problems that might result from the fall in income and asset values.

Farm interest rates should move such that they follow market rates. How closely they follow would depend on recent changes in farm prices. If farm prices are stable, then farm rates should equal market rates. The connection between farm interest rates, market rates, and farm prices should be a well-understood rule written into law. The farm credit system should be self-financing. It should also offer only adjustable rate loans with penalties for early payment, in order to prevent larger farmers, with easier access to other credit sources, from taking advantage of periods when subsidies exist.

(4) Flexible Storage Rules. Just as the proposed credit policy confronts the instability generated in the larger economic system, a program of flexible storage rules would counter the instability generated within the agricultural economy. Current commodity storage programs are intended to moderate price and supply fluctuations by the accumulation of government stocks or by the subsidization of private storers. Traditional stabilization tools, however, have been notoriously unresponsive to market signals, burdening farm policy with costly and sometimes embarrassing stock levels. We concur with several circulating proposals in establishing a program that would respond to market signals and vary with the economic environment.

When a storage program's instruments are left unadjusted as the economic environment changes, they are no longer appropriate for, and are perhaps contrary to, the original policy objectives. This leads to a crisis when dissatisfaction with either program benefits or costs is so widespread that a change in policy becomes inevitable. A flexible storage program, however, would include instruments that change as the economy changes and, therefore, help avoid drastic shifts in policy. If the 1981 Farm Bill had included flexible storage programs, for instance, loan rates could have been conditioned on variables such as stock levels or prices. If loan rates had been a known function of prices and stocks, rates would have declined as world prices fell in the early 1980s. The government would have avoided the buildup in stocks and the disorienting effect of a major change in policy (PIK).

We propose a scheme similar to that of Just and Rausser. The government would determine some target level of stocks based on Treasury costs and the need for safe reserves to meet the goal of food security. The government would also determine a target price and buy or sell a certain amount of a commodity for every 1 percent (say) decrease or increase in price around this target level. Conditioning the target price on the level of stocks would reflect changes in the economic environment and avoid unmanageable divergence between world prices and target levels.

When designing new programs, there is a tendency to leave ample discretion in the case where programs do not respond to changes in the economy in the anticipated manner. Discretion, however, must be minimized in a flexible policy scenario, or political failure in implementation of programs is likely to result. Devising appropriate adjustment rules would be a formidable task, requiring legislative attention to current economic conditions and also to future exigencies. In addition to large initial setup cost, administrative cost may be high because variables in the economic environment would need to be closely monitored. Needless to say, storage costs would be more closely contained and easier to predict.

This proposal would certainly deal with the problems of instability and food security, but it also would tend to promote greater efficiency in agricultural production. A flexible policy would not completely isolate farmers from price changes. Instead, it would encourage farmers to respond to market signals (become more adaptable) because the government would not be promising total insulation of the farm economy which it ultimately cannot provide.

(5) Conservation and Environmental Programs. Issues of agricultural conservation and the environment will continue to grow in importance. Our proposals for conservation and the environment reflect a long-run perspective. Several of the proposals already discussed would aid in achieving social goals of conservation and environmental protection. Targeted income-deficiency payments are meant to encourage smaller farming operations or at least slow the growth of large, capital-intensive farms. An anticyclical credit program would help reduce the incentives for intensive land use that arise during periods when farmers, in a survival mode, seek to increase their immediate cash incomes to deal with credit problems. Finally, perhaps most significantly, the elimination of commodity programs that connect income supports to production levels would ease the extensive and intensive use of land and other resources in production.

We have several additional proposals with greater specificity to the problems of conservation and the environment. First, we propose that the current incentives available for long-run retirement of erodible land be increased. This should be a long-term project that is managed less by traditional bureaucratic authority over agriculture and more by agencies further removed from immediate farm concerns. This would help assure that conservation goals are not forgotten or downplayed in the political process focused mainly on problems related to the profitability of farming.

We also propose to offer incentives to decrease runoff problems and chemical use. The government, again through agencies not immediately concerned with agricultural production, would give investment credits or subsidies to farmers who set up systems that reduce environmental problems. This could be done in much the same way as we have encouraged energy conservation, solar and other alternative energy sources, and the like. This would, however, have to be a recognized long-run program. These subsidies could be funded, in part, by an additional environmental program to tax fertilizer, pesticide, and other chemical use. After all, these productive resources are the source of negative environmental externalities.

All programs, changes in programs, and proposed programs should be subject to conservation and environmental impact studies. A measure of this type would serve to bring greater public attention to the nonfarm effects of agricultural policy instruments. It would increase the difficulty for small groups to bring about policy changes that have limited, direct consequences but far-ranging, indirect effects.

Finally, we propose that the government support more intensive research into conservation and environmentally safe programs and systems that farmers might find profitable to implement. We think it is particularly important to examine water use from common-property resources such as federal water projects and aquifers. Water projects tend to promote production and more intensive use of resources. How can we best design project management, price water, or otherwise limit the long-range conservation and environmental effects of federal water policy? Similarly, we must structure tax or subsidy schemes to reduce the common-property problems associated with drawing water from depletion-troubled aquifers.

(6) Cooperative Export Subsidies (Taxes). We now turn to the most innovative of our proposals--that of cooperative export subsidies (taxes). It is sure to elicit the sharpest criticism from our colleagues in the agricultural economics profession due to its greater, not lesser, emphasis on federal involvement in international commodity trade. We propose to establish an agreement between certain commodity-producing countries to subsidize or tax, on a case-by-case basis, the export of commodities. This is in part to make even more costly the subsidization that goes on routinely by countries wishing to transfer wealth to their own agricultural sectors.

Our suggestion is to manage, to discourage, and to prevent the political failures in other nations that indirectly affect more efficient commodity producers. The EEC spends vast sums subsidizing its agricultural producers, drawing out supplies that would otherwise not be there, and depressing world prices. European consumers and taxpayers carry the immediate burden of such policies, but the farmers in other nations who lose export markets also suffer.

This is a proposal for a collusive strategy that would create incentives to change the nature of political failure in foreign governments. Yes, it is a form of economic retaliation, much as Saudi Arabia's increased production early in 1986 punished non-OPEC members, defectors, and anyone who invested in energy-related projects when oil prices were high. Many individuals and countries will think twice about taking advantage of cartel-supported prices in the future to expand their own production. But while Saudi Arabia may or may not be making its competitors bear a cost today so that higher prices are monopolistically engineered tomorrow, we propose retaliation so that we may be closer to free international trade in years to come.

Some have proposed a grain cartel to increase world prices, claiming that the importing policies (quotas and tariffs) of some countries restrict free trade and lower prices.[21] These proposals, unfortunately, neglect the supply response problem, which must be effectively managed along with the demand problem. We must be assured that, if we do peg a "fair" price for world trade, there is no defection or significant supply response from nonmembers. We must make a credible threat that we will not tolerate exports from countries that have no business in international markets. This is not to say that we should attempt to keep out countries that can efficiently compete with already established producers. As national economies develop, we may even find ourselves in the ranks of the inefficient; but let that be a matter of freely trading markets rather than the result of low prices generated by countries' domestic politics.

We propose that the cooperative export subsidy work in the following fashion. Each country in the cooperative effort contribute to a subsidy fund in proportion roughly equal to its share of the groups' exports. On a commodity-by-commodity, period-by-period basis, a sufficiently low world market price is targeted; and cooperating countries, through subsidies, expand exports. They would do so keeping their export shares constant and maintaining the targeted world price. The cooperative effort would continue matching subsidy after subsidy with offending nations until the inefficient producers adjust their domestic agricultural policies. The cooperative effort would, in effect, underbid every nonmember, gaining a greater share in the world market. Membership would be restricted to those countries that would survive as exporters if all export subsidies were ended.

In some ways, the United States, with the dropping of its loan rates, is unilaterally doing something similar today; but we are also hurting the Canadians and others who, although competitors, are not the offenders. Today's environment, under the 1985 Farm Bill, is much more conducive to forming such a cooperative effort. Once a cooperative effort was established and successful at dealing with the supply problem, then the demand problem could be confronted. A cartel-like framework would have been formalized and test run under conditions that would not have eroded its cohesion; and potential competitors would have been credibly warned against investing in the bureaucratic, political, and economic resources necessary to challenge the cooperative group in the export market. In effect, the cooperative export group would have established the credibility for effective strategic behavior. The group could also manage supply responses to offset the price-reducing monopsonistic practices of many importing countries. These latter practices are also the direct result of political failure which is often motivated by the desire to effectively protect whatever domestic food production occurs. To counter these practices, a cooperative export tax would be introduced. This, too, essentially increases the cost of these political failures.

4. Concluding Remarks

The political feasibility of the proposed set of programs can be seriously questioned. In fact, after reviewing this proposal, public policy students at Berkeley argued strongly that the proposal would never be adopted by the political process. There can be little doubt that the chances are small of implementing a program that involves curtailing property rights established over the years through existing commodity programs. Nevertheless, as argued elsewhere, policy disequilibriums do arise and often lead to major crises (Rausser). When major crises occur, a number of possibilities exist.

Under the 1985 Food Security Act, policy

disequilibriums are very likely. This is, in large part, due to surprises in budget expenditures needed to support various provisions of the 1985 act. If the costs of the 1985 act exceed (by significant amounts) current expected levels of expenditures, a major crisis may be precipitated. In this event, it is important to be positioned with a well-designed, coherent policy program. In the face of major policy disequilibrium, a well-designed policy program must be waiting in the wings to have any chance whatsoever of adoption.

Interest groups would most certainly oppose the design that has been advanced here. However, in a crisis, their opposition will prove ineffective. On a more optimistic note, there is also some hope for success analogous to the recent experience on tax reform. Who would have predicted the success of the U.S. Senate Finance Committee's proposed revisions of the U.S. tax code?

Ultimately, what we desire is a long-term, stable public policy whose transfers (both income and wealth) are made explicit. Unfortunately, current U.S. agricultural policy disguises these transfers. To be sure, this is what interest groups and public officials prefer; the general public finds it more difficult to understand transfers that are implicit and well disguised. To alleviate this problem, we have attempted to design a set of policies which make all transfers explicit. If implemented, informed citizens and potential opponents may then more efficiently counteract such policies.

FOOTNOTES

Much of what appears in the second section may be found in Calvin, Foster, and Rausser (1984) and in Rausser and Foster (1985). The authors wish to thank George Horwich for his comments and editorial suggestions.

1 For four additional treatments of goals of agricultural policy, see Calvin, Foster, and Rausser; Gardner; Knutson (1984b); and Paarlberg (1984b).

2 LeBlanc and Hrubovcak examine tax policy and the effects on agricultural investment. They conclude that a significant share (20 percent) of investment in farm assets can be attributed to tax policy; see also Harl.

3 For good reviews, see Sumner, National Agricultural Forum, and American Farmland Trust.

4 For instance, see Zilberman and Carter and also Hefferman. Zilberman and Carter note that land diversion programs generally advance larger farms, while price and income supports may favor midsize farms. This last point is empirically supported, in the case of Texas, by Smith, Richardson, and Knutson.

5 For a study on this topic, see Tweeten.

6 For a review with emphasis on current policy, see Stucker and Collins.

7 A discussion of biotechnology research and grain production is found in Duvick.

8 Schuh (1984) elaborates this point.

9 For a review of institutional arrangements in world commodity trading, see the Congressional Budget Office report (June, 1983). For an outline of the changing nature of agricultural trade relationships between the United States and other countries, see Josling.

10 For three reviews of this issue, see Batie; Benbrook, Crosson, and Ogg; and Farrell, Sanderson, and Vo.

11 The term "excessive" refers to erosion levels significantly higher than that which soil scientists estimate would maintain long-term productivity (1-5 tons per acre per year, depending upon soil type). For further discussion, see Osteen.

12 Kneese discusses the particular case of the Ogallala aquifer.

13 For studies on the specific commodities discussed in this section, see Babb (dairy policy); Schmitz, Allen, and Leu (sugar); Ray, Tweeten, and Trapp; and Hoover and Sumner (tobacco). The Economic Research Service of the U. S. Department of Agriculture has published several good Commodity Backgrounds for 1985 farm legislation.

14 Overviews of various interest groups are found in the National Agricultural Forum's report on farm policy, Abel and Daft, and Knutson (1984a).

15 The Fertilizer Institute, for example, in preparation of the 1985 legislation, developed their first policy proposal in their 102-year history (The Wall Street Journal, November 23, 1984, p. A9).

16 For a valuable and thorough presentation of alternative policy tools, see Knutson and Richardson.

17 Pasour discusses the free market and farm problems.

18 Petzel reviews futures, options, and comparable

insurance schemes as substitutes for commodity programs.

19 Just and also Just and Rausser analyze automatic adjustment rules, or conditional program instruments, with respect to agricultural policy.

20 For a discussion of these and other issues surrounding food aid, see Paarlberg (1984a).

21 Schmitz et al. detail the issues, problems, and benefits of a grain export cartel. Schuh (1985), on the other hand, dismisses the idea.

REFERENCES

Abel, M., and L. Daft. "Future Directions for U.S. Agricultural Policy." Final Report of Agriculture, Stability, and Growth Conference sponsored by the Curry Foundation, Washington, D.C., 1984.

American Farmland Trust. Future Directions for American Agriculture. Washington, D. C.: American Farmland Trust, 1984.

Babb, E. M. "Dairy." Alternative Agricultural and Food Policies and the 1985 Farm Bill, ed. G. C. Rausser and K. R. Farrell. Giannini Foundation of Agricultural Economics, University of California, Berkeley, and Resources for the Future, Washington, D. C. San Leandro, California: Blaco Publishers, 1984.

Batie, S. "Natural Resource Management and the Future of North American Agriculture." Paper presented at a seminar, "The Future of the North American Granary," sponsored by the Hubert H. Humphrey Institute of Public Affairs, University of Minnesota, June, 1984.

Benbrook, C. M., P. R. Crosson, and C. Ogg. "Resource Dimensions of Agricultural Policy." Alternative Agricultural and Food Policies and the 1985 Farm Bill, ed. G. C. Rausser and K. R. Farrell. Giannini Foundation of Agricultural Economics, University of California, Berkeley, and Resources for the Future, Washington, D. C. San Leandro, California: Blaco Publishers, 1984.

Calvin, L., W. E. Foster, and G. C. Rausser. "Review and Assessment of Alternative Agricultural Policy Proposals." Alternative Agricultural and Food Policies and the 1985 Farm Bill, ed. G. C. Rausser and K. R. Farrell. Giannini Foundation of Agricultural Economics, University of California, Berkeley, and Resources for the Future, Washington, D.C. San Leandro, California: Blaco Publishers, 1984.

Congressional Budget Office. "Agricultural Export Markets and the Potential Effects of Export Subsidies." Staff working paper, Washington, D.C., June, 1983.

________. Farm Revenue Insurance: An Alternative Risk-Management Option for Crop Farmers. Washington, D.C.: U.S. Government Printing Office, August, 1983.

Duvick, D. "North American Grain Production-Biotechnology Research and the Private Sector." Paper presented at a seminar, "The Future of the North American Granary," sponsored by the Hubert H. Humphrey Institute of Public Affairs, University of Minnesota, June, 1984.

Farrell, K. R., F. H. Sanderson, and T. Vo. "Feeding a Hungry World." Resources, No. 76, Resources for the Future, Washington, D. C., 1984.

Gardner, B. "Domestic Policy Options for the Future of U.S. Agriculture." Paper presented at "Agriculture, Stability, and Growth: Toward a Cooperative Approach," a conference sponsored by the Curry Foundation, Washington, D. C., February, 1984.

Harl, N. "Impact of Tax Policy on American Agriculture." United States Agricultural Policy for 1985 and Beyond, ed. J. S. Hillman. Tucson: University of Arizona, 1984.

Hefferman, W. "Examining the Consequences of Recent Agricultural Policy on Farm Families in Rural Communities." Restructuring Policy for Agriculture: Some Alternative, ed. S. Batie and J. P. Marshall. Blacksburg: Virginia Polytechnic Institute and State University, 1984.

Hoover, D. N., and D. Sumner. "Tobacco and Peanuts." Alternative Agricultural and Food Policies and the 1985 Farm Bill, ed. G. C. Rausser and K. R. Farrell. Giannini Foundation of Agricultural Economics, University of 8alifornia, Berkeley, and Resources for the Future, Washington, D.C. San Leandro, California: Blaco Publishers, 1984.

Josling, T. "Agricultural Trade Among Friends: The Parlous State of U.S. Trade Relationships with the Industrial West." United States Agricultural Policy for 1985 and Beyond, ed. J. S. Hillman. Tucson: University of Arizona, 1984.

Just, R. E. "Automatic Adjustment Rules in Commodity Programs." U.S. Agricultural Policy: The 1985 Farm Legislation, ed. B. L. Gardner. Washington, D. C.: American Enterprise Institute, 1985.

Just, R. E., and G. C. Rausser. "Uncertain Economic Environments and Conditional Policies." Alternative Agricultural and Food Policies and the 1985 Farm Bill, ed. G. C. Rausser and K. R. Farrell. Giannini Foundation of Agricultural Economics, University of California, Berkeley, and Resources for the Future, Washington, D.C. San Leandro, California: Blaco Publishers, 1984.

Kneese, A. V. "Water Resource Constraints: The Case of the Ogallala Aquifer." Paper presented at a seminar, "The Future of the North American Granary," sponsored by the Hubert H. Humphrey Institute of Public Affairs, University of Minnesota, June, 1984.

Knutson, R. "The Goals of Agriculture and Food Policy." Paper prepared for the American Enterprise Institute's Agricultural Studies Project, December, 1984a.

_________. "The Public Interest in Agricultural Policy." United States Agricultural Policy for 1985 and Beyond, ed. J. S. Hillman. Tucson: University of Arizona, 1984b.

_________ and J. W. Richardson. "Alternative Policy Tools for U.S. Agriculture." Agricultural and Food Policy Center, Texas A&M University, August, 1984.

LeBlanc, M., and J. Hrubovcak. "The Effect of Tax Policy on Aggregate Agricultural Investment." American Journal of Agricultural Economics. (To be published in the November, 1986, issue.)

National Agricultural Forum Domestic Policy-Alternatives Task Force. Alternatives for U.S. Food and Agricultural Policy. Washington, D.C.: National Agricultural Forum, 1984.

Osteen, C. "Impacts of Farm Policies on Soil Erosion." U.S. Economic Research Service, Natural Resource Economics Division, ERS Staff Report No. AGES841109, January, 1985.

Paarlberg, D. "U.S. International Agricultural Policy." Options Paper prepared for Center for National Policy, January, 1984a.

_________. "Purposes of Farm Policy." Paper prepared for the American Enterprise Institute's Agricultural Studies Project, December, 1984b.

Pasour, E. C., Jr. "The Free Market Answer to U. S. Farm Problems." The Backgrounder, No. 339. Washington, D.C.: The Heritage Foundation, 1984.

Petzel, T. "Toward a Market Orientation: The Dilemma Facing Farm Policy in the 1980s." U.S. Agricultural Policy: The 1985 Farm Legislation, ed. B. L. Gardner. Washington, D. C.: American Enterprise Institute, 1985.

Rausser, Gordon C. "Political Economic Markets: PESTs and PERTs in Food and Agriculture." American Journal of Agricultural Economics, Vol. 64, No. 5 (December, 1982), pp. 821-833.

Rausser, G. C., and W. E. Foster. "A Synthesis of Major Studies and Options for 1985." The Dilemmas of Choice, ed. K. A. Price. Washington, D. C.: The National Center for Food and Agricultural Policy, Resources for the Future, Inc., 1985.

Ray, D. E., L. G. Tweeten, and J. N. Trapp. "Linkages to the Livestock Sector." Alternative Agricultural and Food Policies and the 1985 Farm Bill, ed. G. C. Rausser and K. R. Farrell. Giannini Foundation of Agricultural Economics, University of California, Berkeley, and Resources for the Future, Washington, D.C. San Leandro, California: Blaco Publishers, 1984.

Schmitz, A., R. Allen, and G. J. M. Leu. "The U.S. Sugar Program and Its Effects." Alternative Agricultural and Food Policies and the 1985 Farm Bill, ed. G. C. Rausser and K. R. Farrell. Giannini Foundation of Agricultural Economics, University of California, Berkeley, and Resources for the Future, Washington, D.C. San Leandro, California: Blaco Publishers, 1984.

__________, A. F. McCalla, D. O. Mitchell, and C. A. Carter. Grain Export Cartels. Cambridge: Ballinger Publishing Company, 1981.

Schuh, G. E. "Trade and Macroeconomic Dimensions of Agricultural Policies." Alternative Agricultural and Food Policies and the 1985 Farm Bill, ed. G. C. Rausser and K. R. Farrell. Giannini Foundation of Agricultural Economics, University of California, Berkeley, and Resources for the Future, Washington, D.C. San Leandro, California: Blaco Publishers, 1984.

__________. "Improving U. S. Agricultural Trade." The Dilemmas of Choice, ed. K. A. Price. Washington, D.C.: The National Center for Food and Agricultural Policy, Resources for the Future, Inc., 1985.

Smith, E., J. Richardson, and R. Knutson. "Impact of Farm Policy on the Structure of Agriculture in the Texas Southern High Plains." Agricultural and Food Policy Center, Texas A&M University, 1984.

Stucker, Barbara C., and Keith J. Collins. The Food Security Act of 1985: Major Provisions Affecting Commodities. U.S. Economic Research Service, Agriculture Information Bulletin No. 497, January, 1986.

Sumner, D. "Farm Programs and Structural Issues." U.S. Agricultural Policy: the 1985 Farm Legislation, ed. B. L. Gardner. Washington, D.C.: American Enterprise Institute, 1985.

Tweeten, L. "Farm Financial Stress, Structure of Agriculture, and Public Policy." U.S. Agricultural Policy: The 1985 Farm Legislation, ed. B. L. Gardner. Washington, D.C.: American Enterprise Institute, 1985.

U.S. Economic Research Service. Commodity Backgrounds for 1985 Farm Legislation, Agricultural Information Bulletin No. 465-478, 1984.

Zilberman, D., and H. O. Carter. "Structural Dimensions of Agricultural Poli cies." Alternative Agricultural and Food Policies and the 1985 Farm Bill, ed. G. C. Rausser and K. R. Farrell. Giannini Foundation of Agricultural Economics, University of California, Berkeley, and Resources for the Future, Washington, D.C. San Leandro, California: Blaco Publishers, 1984.

TABLE 1

Intended Objectives of Proposed Policies

	Transfer wealth to farmers	Reduce agriculture risk	Assure safe level of food stocks	Assure reasonable food prices to consumers	Make farm sector flexible	Protect family farms	Conserve natural resources	Minimize public costs	Avoid policy crises
Free markets				x	x			x	x
Reinforce markets:									
revenue insurance		x		x	x			x	
aid to forward mkts		x		x	x			x	
flexible storage		x	x	x	x			x	x
Supply contraction:									
voluntary controls	x						x		
mandatory controls	x						x	x	
quota/excess production tax	x						x	x	
Demand expansion:									
food stamps	x			x					
export enhancement	x								
Direct income aid:									
negative income tax	x	x			x	x			x
sector income guarantees	x	x							
deficiency payments	x	x		x					
Stock accumulation:									
FOR/CCC loans	x	x	x	x					
flexible loans		x	x	x				x	x
direct purchase	x	x	x						

11

The Prospects for Successfully Restructuring the Farm Credit System

Freddie L. Barnard and William D. Dobson

As is well known, many U.S. farmers have experienced financial problems in recent years. Harrington found that about one-third of U.S. commercial farmers faced serious financial problems, severe financial problems, or technical insolvency in 1985. Lee indicated that about 10 percent of the Nation's farmers were so highly leveraged in 1986 that they probably would not survive. The financial troubles of the farmer have created difficulties for agricultural lenders, including the Farm Credit System (FCS), banks, individuals, insurance companies, and the Farmers Home Administration. The problems are most acute for lenders such as the Federal Land Banks which have parts of their loan portfolios in undercollateralized farmland mortgages. The purposes of this article are to examine the origins of the problems facing the FCS, the effectiveness with which the 1985 and 1987 federal assistance programs for the FCS achieved short-term objectives, and prospects for successful evolution of this major agricultural lending organization during the next several years. Emphasis is placed on

---

This chapter includes a discussion of certain major developments affecting the Farm Credit System. An article containing some material in this chapter appeared in Freddie L. Barnard and W.D. Dobson, "The Problems and Prospects of the Farm Credit System," Agribusiness: An International Journal, Vol. III, No. 3 (1987), pp. 323--337.

identifying, and examining the implications of, management and policy decisions which contributed to problems of the FCS and solutions to those problems.

As of January 6, 1988, the FCS consisted of 12 Federal Land Banks (FLBs) which provide credit to farmers for land and capital purchases through local Federal Land Bank Associations (FLBAs), 12 Federal Intermediate Credit Banks (FICBs) which operate through Production Credit Associations (PCAs) to provide farmers with production credit--e.g., for purchase of feed, seed, and fertilizer --and intermediate credit, and 13 Banks for Cooperatives (BCs) which provide seasonal and term loans for agricultural Cooperatives. The FCS is currently regulated by the Farm Credit Administration (FCA).

Unlike other private lenders, the FCS is restricted to making loans only to farmers, fishermen, agricultural Cooperatives, and rural residents who meet eligibility standards set by law. On December 31, 1987, the FLBs/FLBAs, FICBs/PCAs, and BCs, respectively, had about $32 billion, $9 billion, and $8 billion in net loans outstanding. These organizations obtain funds for lending mainly through the sale of securities in financial markets which are the joint obligation of the 37 farm credit banks in the FCS, stock purchased by borrowers, surpluses, and retained earnings.

Origins of the Problems of the Farm Credit System

How did the financial problems of the FCS arise? These problems arose partly because economic conditions affecting farming became substantially worse than FCS officials expected and the ability of farmers to repay loans and employ financial leverage to advantage declined by more than FCS officials anticipated. This, in turn, placed unanticipated stresses on the FCS, caused the loan volume of the FCS to contract, and caused the System to incur losses beginning in 1985. The basis for these findings emerge partly from comparing results of the FCS Project 1995 study to actual events.

Implications of the Project 1995 Study

The major Project 1995 study (released in June, 1984), which was directed by FCS officials, contains forecasts supplied by university professors, business consultants, and persons employed by the FCS of economic conditions that the FCS might face during 1985-1995.

Seven forecasts appearing in the Project 1995 study for 1985-1995 relating to the economic environment and the amount of credit that would be used by farmers are summarized in Table 1 (Tables at the end of chapter). Actual values for these variables for 1973-81, 1982-84, 1985, 1986, and 1987 are included for purposes of comparison and use in analyses appearing below.

The Project 1995 forecasts in Table 1--which are part of a standard scenario--describe a U.S. economy which during approximately the next decade would perform more favorably in terms of real Gross National Product (GNP) growth and employment than during either the generally inflationary years of 1973-81 or the 1982-84 period. Indeed, the real GNP growth and unemployment rate forecasts appearing in Table 1 are 22 percent higher and 32 percent lower, respectively, than the average actual values recorded for these variables during 1973-84.

An excessively optimistic forecast of real GNP growth would contribute to overestimates of the demand for U.S. farm products. It has been estimated that a one percent increase in real GNP results in about a 0.4 percent increase in farm level demand in the U.S. Hence, if real GNP grew at 3.2 percent per year as forecast in the Project 1995 study, then U.S. farm level demand would grow about 1.28 percent per year. If, on the other hand, real GNP grew only at the 2.5 percent average rate recorded during 1973-84, farm level demand would grow only 1.00 percent per year or at about a 22 percent slower rate. This lower rate of increase in the domestic demand for farm products, together with smaller than expected agricultural exports, produced a weaker farm economy than anticipated in the Project 1995 study.

Forecast No. 3 in Table 1 is a conditional forecast which states that the prime interest rate will drop to 8 percent or 9 percent "if federal budgets are brought under control". The U.S. Supreme Court found that certain provisions of the Balanced Budget and Emergency Deficit Control Act of 1985 (Gramm-Rudman-Hollings legislation) were invalid in a July 7, 1986 decision. This court decision adds to the uncertainty regarding whether the important deficit control condition associated with this forecast will be met. However, since the prime rate for 1986 and 1987 was 8.3 percent and 8.2 percent, respectively, and the prime interest rate remained at 9 percent or below through at least June 1988, the Project 1995 forecast regarding prime interest rates was a reasonably accurate forecast for the last half of 1986,

1987, and the first half of 1988.

Forecast No. 4 in Table 1 suggests that domestic inflation rates in the U.S. will average 5 percent to 7 percent during 1985-1995. Several factors, including weak energy prices, wage settlements that are low by standards of the 1970s, and adequate manufacturing capacity appear likely to keep inflation substantially below the levels of 1973-81 through mid 1988 at least. Thus, it is unlikely that, in the near future, inflation will cause a flight of dollars from financial assets to inflation hedge assets such as farmland. If this reasoning is correct, it has obvious implications regarding farmland prices, the value of farmland as loan collateral and FLB loans.

The authors of the Project 1995 study forecasted a slowdown in the rate of growth of U.S. agricultural exports (Forecast No. 5, Table 1) rather than a sharp decline. U.S. exports of agricultural products were buoyed by a relatively weak dollar and strong demand in foreign markets during parts of 1973-81. After this period, U.S. exports of agricultural products declined, reaching levels in fiscal 1986 which were 40 percent below those recorded in fiscal 1981. The declines in the exchange value of the dollar during 1985 through mid-1988 and use of the the lower loan rates and export subsidies authorized by the Food Security Act of 1985 helped to increase U.S. agricultural exports. Thus, ironically, the forecast appearing in the Project 1995 study which suggests that "U.S. agricultural exports are expected to grow but not at the high rates of the 1970s" was correct for 1987 and may be correct for additional years in the period. However, such growth will occur from a lower base than anticipated by the authors of the Project 1995 study, leaving more excess capacity in farming than anticipated in the study.

Farm real estate debt was forecast in the Project 1995 study to exhibit a nominal growth of 5.2 percent per year during 1985-95 and, given the inflation rate forecast (5 percent to 7 percent), would exhibit slow real growth or reductions in real growth in these years. Nonreal estate debt of farmers was forecast to increase at rates below those recorded during 1973-81 but above those recorded during 1982-84.

The forecasts regarding credit are certain to overstate the actual amounts of credit used by farmers for at least the first few years of the 1985-95 period, since after peaking in 1982 at about $217 billion, the amount of credit used by U.S. farmers has declined. FCS net farm loans outstanding in the U.S. declined by 39 percent from $67 billion on December 31, 1984 to $41 billion on December 31, 1987. In view of the gap that has existed during most of the 1980s between farm interest rates and typical yields on farm assets, farmers will have a strong incentive to reduce debt further; moreover history suggests that farmers remain cautious about using credit for an extended time after it has again become profitable for them to use financial leverage. If these forecasts regarding financial leverage are correct, FCS officials face the task of managing an organization whose loan volume will continue to shrink in the years ahead.

Thus, officials of the FCS encountered in the mid-1980s a harsher economic environment than anticipated in the Project 1995 planning study. FCS officials might be criticized for accepting forecasts of real GNP which were more favorable than those witnessed during 1973-84. However, even this indictment may be too strong since the 3.2 percent real GNP growth figure was similar to the one being distributed to other U.S. businesses in 1984 by the 42 economists contributing to Blue Chip Economic Indicators who forecasted in March 1984 and October 1984 that growth of real GNP would average 3.1 percent during 1985-1994. There is no way to assess meaningfully whether other changes in economic conditions should have been anticipated more accurately by the FCS. However, the comparison of forecasts in the Project 1995 study and actual events does serve as a reminder of how difficult it is for firms to obtain accurate long-term economic forecasts. The disparity between forecasts and events also suggests that the FCS would benefit from employing management strategies, notably diversification of portfolio, that would allow the System to deal effectively with a wider range of economic environments.

Clearly, harsh and partially unforeseen economic conditions put about one-third of U.S. farmers in debt categories which, as noted earlier, caused them to have financial problems in the mid-1980s. These developments, in turn, produced losses and other financial problems for the FCS and other agricultural lenders.

Lending Decisions of the FCS, Commercial Banks, and Insurance Companies

Did the FCS finance borrowers who were poorer credit risks or, for other reasons, leave itself more vulnerable to financial difficulties than other private agricultural lenders? Information is presented below on this question as it relates to decisions of commercial banks, insurance companies, and the FCS.

In January 1985, 59 percent of outstanding FCS farm loans were in the hands of farmers with debt-asset ratios exceeding 40 percent. In this same month, an identical 59 percent of the outstanding farm loans of commercial banks consisted of debts of farmers with debt-asset ratios exceeding 40 percent. Farmers with debt-asset ratios above 40 percent may have serious financial problems, severe financial problems, or be technically insolvent. Since the percentage of farm loans in the hands of farmers with debt-asset ratios exceeding 40 percent was equal for both borrower groups, this comparison suggests that, as a group, FCS borrowers were neither more nor less creditworthy than farmers who borrowed from commercial banks.

Irwin makes a related point, noting that PCAs have slightly less than half of the loan volume that commercial banks have in agriculture, and they recently have had about half as much in loan charge-offs. This, he argues, indicates that any charges of credit mismanagement logically must be levied at both or at neither.

While these comparisons suggest that FCS lenders were neither more nor less able to avoid problem farm loans than commercial banks, the problems of the FCS are more serious because of the higher percentage of farm loans in the portfolio of FCS lenders. In addition, the problems of the FCS are more severe because of the larger percentage of undercollateralized farm mortgage debt in the FCS portfolio and because the FCS has used a higher percentage of its loans to finance larger farmers than have commercial banks. Regarding the latter point, in January 1985, 41 percent of the outstanding farm debt of the FCS consisted of loans made to farmers with annual sales of $250 thousand to $500 thousand and over $500 thousand. The comparable figure for commercial banks was 33 percent. This difference is noteworthy since larger, commercial farmers have experienced more severe financial problems in the mid-1980s than smaller, part-time farmers.

The interest rates charged by FLBAs and PCAs were relatively low during 1980-85. For example, national average interest rates charged by PCAs were about two percentage points less than those charged by commercial banks on similar loans during 1980-85. U.S. average FLBA interest rates for farmland
mortgages also were about two percentage points lower than

those charged by insurance companies during 1980-85, although FLBA interest rates did rise above those charged by the life insurance companies late in 1985. The relatively low interest rates charged by the FLBAs helped the organization to increase its share of U.S. farm real estate mortgages by about 8 percentage points during 1980-85. Equipped with hindsight, one can ask whether the interest rates charged by the FLBAs, in particular, during 1980-85 should have been higher. Higher rates would have reduced the FLBAs market share, but they also would have discouraged some FCS borrowers from obtaining loans which they could not repay and could have produced FCS reserves that were more adequate to deal with the harsh economic environment encountered in the mid-1980s.

Federal Assistance for the FCS in 1985

Selected events leading to and following federal action in December 1985 to assist the FCS--the largest rescue of a U.S. financial institution in history--are chronicled in Table 2 (Tables at the end of chapter). One interpretation of the events described in Table 2 is that from early 1985 until about September, 1985 there were problems in certain FCS districts (notably Spokane and Omaha), but these problems were not important enough to push the interest rate spread on FCS securities sold in securities markets much more than 20 basis points above comparable U.S. Treasury securities. Then as large crops became a near certainty for 1985, export markets remained weak and the serious nature of the problems within the FCS became more apparent, the interest rate spreads between FCS securities and Treasury issues widened. Ultimately, the problems of the System led officials of the FCS and the FCA to request federal aid which was granted in the form of the legislation signed into law on December 23, 1985.

Objectives of the 1985 Federal Legislation

Two immediate objectives of the new legislation, which were to reassure financial markets of the soundness of FCS securities and keep FCS farmer customers from withdrawing about $5 billion in FCS stock they held and taking sound loan business elsewhere, were to be achieved partly through provisions in the legislation which would:

1. Give the FCS broad authority to use its own resources to bolster financially weak units

within the System.

2. Make the FCA a stronger, arms-length regulator of the FCS.
3. Give the U.S. Treasury authority to purchase obligations of a rechartered unit called the Farm Credit System Capital Corporation (FCSCC). These purchases could be made only after the FCA had certified that the FCS needed financial aid and had made the maximum practical effort to deal with the financial stress by using its own resources. The Congress would appropriate any funds used by the Treasury to purchase obligations of the FCSCC.
4. Require the FCS to provide more financial information to borrowers and stockholders.

The FCSCC mentioned in provision (3) would provide a central source of financial help to individual units of the FCS through purchases of stock from, or loans and contributions to, local units which could not operate without financial aid. To finance its activities the FCSCC could draw funds from the strongest System institutions by steps, including those which would require stock purchases or assessments made under guidelines set by the FCA. The power to draw on resources of other FCS units would be limited by provisions stating that the FCSCC could not require other units to employ the stock held by member borrowers as a contribution to the new central fund and the FCSCC's levies could not rise to the point that the financial viability of local institutions would be imperiled or local institutions would be unable to make credit available to borrowers on reasonable terms.

Effectiveness of the 1985 Federal Legislation

The effectiveness of the legislation is assessed below for time periods designated as Period I and Period II. In terms of specific dates, Period I began in December, 1985 when passage of the new legislation became imminent and ended in March-April 1986 after market reactions to the FCS financial report for 1985 had been recorded. Period II began immediately after Period I and lasted until January 6, 1988 when the Agricultural Credit Act of 1987 was signed into law.

Period I: Effects of the law in Period I include those which relate to whether the immediate objectives of the legislation--i.e., to restore the confidence of the financial markets in FCS securities and preserve the value of member stock--were achieved.

Financial market participants apparently became persuaded during December 1985-March 1986 that FCS securities were backed by a workable federal guarantee since the FCS-Treasury interest rate spread declined from the 80+ basis point level recorded in October-December 1985 to about 33 basis points in April 1986 in the face of the $2.7 billion loss recorded by the FCS for 1985 (Table 2). Presumably, the decline in the interest rate spreads reflects mainly the effects of the new federal legislation. However, these developments also may reflect reactions within financial markets to efficiency measures adopted and consolidations and mergers that occurred within the FCS. The merger of 53 FLBAs and 30 PCAs in the Louisville Farm Credit District which was approved by member-borrowers in December 1985 is an example of an action taken both to gain efficiencies in administration of credit programs and to give member stock the financial backing of the district-wide association.

How effectively the federal assistance measure allayed concerns of farmer borrowers about the value of their stock and helped the FCS keep the business of its farmer members cannot be discerned from figures for Period I alone. However, some information on these points is provided by data on changes in the share of total U.S. farm credit provided by FCS lenders during December 31, 1984-December 31, 1985 and changes in total FLB and PCA loans outstanding during the period (Table 3) (Tables at the end of chapter). The percentage of farm real estate debt provided by the FLB declined by 2.1 percentage points from December 31, 1984 to December 31, 1985, while banks, insurance companies and the FmHA increased their market shares. Farm lenders who served on credit task forces with the authors suggested that lower interest rates charged by banks and concern about stock values caused some FLB borrowers to refinance with banks and other lenders during 1985. Total FLB loans outstanding declined by $4 billion (8.2 percent) during December 31, 1984 to December 31, 1985. However, a substantial part of the decline in FLB loan volume apparently represented a paydown by borrowers and loan losses since the decline exceeded the increase in farm mortgage loans of commercial banks and the FmHA. PCAs had about a four percentage point lower share of the farm nonreal estate debt on December 31, 1985 than on December 31, 1984, while the FmHA and the Commodity Credit Corporation increased their shares of the market during this period (Table 3).

Figures obtained from the U.S. Department of Agriculture's Farm Land Market Survey show that Federal Land Banks in the U.S. extended only 25 percent of the credit used to finance farmland transfers during the year ending February 1, 1986, down 6 percentage points from the comparable figure for February 1, 1985. Commercial banks were the biggest gainers increasing their share of the credit used to finance farmland transfers from 13 percent for the year ending February 1, 1985 to 21 percent for the year ending February 1, 1986.

The FLBs and the PCAs clearly lost market share during the periods described above. However, in view of the results obtained in financial markets and the limited exodus of borrowers from the FCS during late 1985 and early 1986 (nothing analogous to a major run on the FCS occurred during this period), the sponsors of the legislation seem justified in claiming that the federal assistance program worked effectively to bolster the System during Period I. These results also were obtained without federal budget outlays.

Period II: The level of confidence of the financial markets in FCS securities, the value of member stock and other measures of the financial health of the FCS were influenced by many things during Period II, including the speed of the recovery of the farm economy, how well the complex assistance plan was administered by FCS and FCA officials, the timeliness with which funds needed for federal assistance of the FCS were provided, the competitiveness of FCS interest rates to member borrowers, the effectiveness of the efficiency measures adopted by the FCS for reducing operating costs and degree of member support.

On the positive side, there was the legacy of success during Period I. In addition, experienced credit officials were appointed to positions on the FCSCC and FCA Board (Table 2). However, as noted in Table 2, problems soon surfaced. Early in 1986, officials of the Amarillo, Texas PCA asked the FCA for permission to withdraw from the FCS to avoid sharing the Association's reserves with financially weaker units of the FCS. The FCA denied this request but the Amarillo PCA then initiated court action in an effort to secure its objectives (Table 2). Officials of the Farm Credit Bank of Springfield, Farm Credit Bank of Texas, and PCAs in several states challenged the legality of the loss-sharing provisions provided by the 1985 federal legislation in the third quarter of 1986, raising constitutionality issues and charging that the levies on financially stronger FCS units were so large that they would damage the solvency of contributing FCS units. It was clear that these and similar actions would render unworkable the loss-sharing provisions of the federal legislation. Secondly, reports received by the authors indicated that FLB interest rates during 1986 had become sufficiently uncompetitive in some FCS districts that loan paydown rates rose to several times normal levels and the FLB market share of new farm real estate loans had fallen to about one-quarter of normal levels. Finally, losses totalling $1.9 billion were reported for the FCS for 1986 (Table 2).

FCS officials sought permission from the FCA in the second quarter of 1986 to lower interest rates to reduce the cost of credit for farm borrowers, help the FCS to maintain market share, and help the organization to maintain its income generating capacity. In mid-1986 the FCA approved requests for lower rates proposed by FCS districts, with the proviso that a district's weighted average interest rate to all borrowers would not decline by more than 0.5 percentage point from current levels. The proviso reflected FCA's concerns that larger interest rate reductions would cause FCS revenues to fall short of those needed to generate the reserves needed for dealing with nonperforming loans and loan charge-offs and hasten the time when federal budget outlays would be needed to bolster the FCS.

Legislation signed into law in October 1986 allowed FCS units to set interest rates without prior approval of the FCS (Table 2). This action gave FCS units discretion to establish interest rates which balanced the need to maintain competitive market shares against the need to charge interest rates that reflect the different risks and costs involved in serving different borrowers and generate adequate reserves.

Numerous forecasts were made of when exhaustion of FCS surpluses would necessitate an infusion of federal dollars into the FCS. For example, a General Accounting Office (GAO) Study presented to a Congressional Committee in September 1986 suggested that FCS surpluses would be exhausted by early 1987. Unofficial FCS and FCA forecasts indicated that FCS surpluses would last until 1988. A third forecast suggested the FCS would exhaust its surpluses before the end of 1987 if losses continued at the rate registered during 1985 and 1986. Finally, while forecasters recognized that the rate of depletion of FCS surpluses could be slowed if FCS units used the regulatory accounting provisions which permit certain losses to be amortized over a 20-year period, some noted that such an action would not necessarily eliminate the need for an infusion of federal dollars for the System.

Federal Assistance for the FCS in 1987

Although the financial position of the FCS appeared to improve during 1987, the System continued to experience financial problems. The reported losses for the first and second quarters were only $155 million and $46 million, respectively; whereas, net incomes of $4 million and $180 million were reported for the third and fourth quarters, respectively. This resulted in a loss for 1987 of only $17 million, which was sharply lower than the $2.7 billion and $1.9 billion losses reported for 1985 and 1986, respectively. However, the major reason for the improved income situation for 1987 compared to 1986 was a $1.994 billion reduction in the provision for loan losses. Also, the loss on other property owned was only $12 million for 1987 compared to $233 million for 1986. At the end of 1987 many of the System's FLBs were still experiencing financial stress, the System continued to carry a large volume of high-cost debt and the System continued to experience a decline in loan volume. This set of circumstances made it difficult for the System to generate the earnings needed to keep financially troubled FCS institutions solvent.

In March 1987, officials of the FCS asked the Federal Government for additional financial assistance. In response, both the House and the Senate began to develop bills to resolve the difficulties of the FCS. The Agricultural Credit Act of 1987 which provided additional assistance for the FCS was signed into law on January 6, 1988.

Objectives of the 1987 Federal Legislation

The Agricultural Credit Act of 1987 amended the Farm Credit Act of 1971, the Federal statute under which the FCS operates. The major provisions of the Act that apply to the FCS establish mechanisms for providing financial assistance to System institutions, protect the value of certain borrower stock establish borrower rights protection to System borrowers, provide for organizational changes in the System, establish a Farm Credit Insurance Corporation and Fund, and create a secondary mortgage market for agricultural loans.

A Farm Credit System Assistance Board is established by the Act to aid System institutions when the value of the stock owned by borrowers falls below par. If borrower stock is impaired by less than 25 percent, the institution may apply to the Board to receive financial assistance. If stock value is below 75 percent of par using generally accepted accounting principles, the institution must apply for assistance. The charter of the Farm Credit System Capital Corporation was revoked by the legislation, and all assets, liabilities, contractual obligations, security, and title instruments were transferred to the Assistance Board.

A Farm Credit System Financial Assistance Corporation is established by the legislation to aid System institutions. To obtain needed funds, the Assistance Corporation is authorized to issue up to $4 billion in 15-year uncollateralized bonds guaranteed by the U.S. Treasury. On March 6, 1988 the Assistance Corporation began issuing Treasury-guaranteed bonds which may total as much as $2.8 billion in value. After January 1, 1989, if the Assistance Board determines that more funds are needed to aid System units, it may issue up to $1.2 billion in additional bonds.

The U.S. Treasury pays all interest costs on each bond for the first five years. During the next five years, the Treasury will pay up to one-half of the interest. The FCS is required to pay all of the interest costs during the last five years. After 15 years the FCA, in consultation with the Treasury, will determine a schedule under which the System will repay the principal and the interest payments that were made by the Treasury.

The Act also protects the stock of FCS borrowers by requiring System institutions to retire stock at par value on loans that are paid in full. This guarantee covers stock that was outstanding on the date of enactment of the Credit Act of 1987, and stock that would be purchased within the earlier of nine months after passage of the 1987 Credit Act or the adoption of a new capitalization plan by the System institutions.

The Act requires all FLBs, PCAs, and other financial institutions that discount with the Federal Intermediate Credit Banks to restructure distressed loans if restructuring is less costly than foreclosure. The Act also prohibits System lenders from foreclosing any loan due to declining collateral value or previous delinquency, if the borrower brings current all accrued payments of loan principal, interest, and penalties. Farmer-borrowers of the FCS also obtained rights of "first refusal" protection, which gives them the option to repurchase or lease foreclosed property they previously owned. Finally, the Act increases borrower access to FCS loan documents and information, and provides for an independent "second opinion" appraisal which may be used during disputes involving collateral values.

A multistep process for restructuring the System is outlined in the legislation. This process includes the required merger of the FLB and FICB in each Farm Credit District within six months of passage of the Credit Act. It also requires the boards of directors of each PCA and FLBA that share substantially the same geographical territory to submit a plan for merging the two associations to a stockholder vote within six months of the district bank merger. The law sets in motion an 18-month process for developing a plan to merge the existing 12 Farm Credit Districts into no fewer than six districts. The law also provided for a committee to develop a proposal for the voluntary merger of the BCs into a National Bank for Cooperatives.

The law establishes a Farm Credit Insurance Corporation and Fund to insure the timely payment of principal and interest on notes, bonds, debentures, and other obligations of eligible and participating FCS institutions. The fund will begin insuring obligations January 5, 1993. Joint and several liability of System institutions for FCS debt obligations is maintained. However, it will only be triggered if all monies in the insurance fund are exhausted.

Finally, the 1987 Act establishes the Federal Agricultural Mortgage Corporation, within the FCS to facilitate the development of a secondary market for agricultural real estate loans. Qualified agricultural real estate mortgages will be sold on a nonrecourse basis by loan originators to certified agricultural mortgage marketing facilities (poolers). The poolers package the loans that serve as collateral for securities purchased by the investing public.

Effectiveness of the 1987 Federal Legislation

At the time of this writing, only six months of history were available on the effects of the 1987 Act. Hence, the effectiveness of the legislation cannot yet be fully assessed. The discussion that follows first addresses the effects of the new legislation that occurred during the six months following enactment. Then some fundamental problems are discussed which need to be addressed if the financial health of the System is to show long-run improvement.

Initial Effects

The effects of three provisions of the new legislation during the first half of 1988 can be assessed. First, the mechanism to provide financial assistance to System institutions was tested in February and again in May when the Assistance Board authorized $25 million and $90 million to aid the Jackson and the Louisville FLBs, respectively (Table 2). In both instances, the mechanism was an effective means of channeling financial assistance to the institutions. However, in May, 1988 the Assistance Board determined that the probable cost of liquidation of the Jackson FLB was substantially less than the probable cost of restoring its viability and competitiveness through financial and other assistance. Consequently, the Assistance Board requested the FCA to appoint a receiver for the FLB of Jackson to liquidate that institution (Table 2).

Second, the provisions of the Act that protect the value of borrower stock were tested during the Spring of 1988. During the first quarter of 1988, six PCAs in liquidation in the Spokane and Omaha districts were granted a total of $20.6 million to permit retirement at par of previously frozen borrower stock. Also, in conjunction with the request by the Assistance Board that the FCA appoint a receiver for the FLB of Jackson, $5 million of assistance was approved to permit the FLB to retire eligible borrower stock at par value.

Finally, on June 30, 1988 eight of the 12 Banks for Cooperatives voted to merge into a single national bank. The new institution will be called the National Bank for Cooperatives, and will have total assets of $9.3 billion. The four Banks for Cooperatives voting to remain independent are based in Springfield, Massachusetts; St. Paul, Minnesota; Jackson, Mississippi; and Spokane, Washington.

The FCS has taken steps to implement other provisions of the 1987 Act. Loan restructuring programs and procedures for disposing of acquired property are being re-evaluated to ensure that the borrowers' rights guaranteed in the 1987 Act are being protected. Efforts are also underway to implement the mandatory merger of

FLBs and FICBs. An assessment of the effectiveness of these and other provisions cannot be made at the time of this writing.

The likelihood of a financial turnaround in 1988 for some FCS banks was rapidly decreasing in mid-1988 because of the drought in much of the middle of the country. The Federal Land Banks in St. Paul, Minnesota and Louisville, Kentucky appear to be the most vulnerable to a drought-induced setback.

Long Term Problems

Certain fundamental problems may need to be addressed more fully by FCS officials if the financial health of the FCS is to show long-run improvement. These include problems relating to diversification of portfolio and management of credit.

While the FCS may extricate itself from current problems with federal assistance, the organization will be vulnerable to financial difficulties when another major downturn in farming occurs. Thus, additional diversification of the FCS portfolio may be needed if such problems are to be reduced. Indeed, the financial performance of the more diversified BCs during 1985 and 1986 argues for greater diversification of the FCS portfolio. How might this be done? FLBAs and PCAs might seek legislation which would permit them to provide additional financial services and make loans to agricultural and nonagricultural businesses in rural areas as well as farm loans. Adding non-agricultural business loans could, of course, change the FCS in fundamental ways and require employees of the FLBAs and PCAs to acquire a broader range of lending skills. As an alternative to diversification, the FCS might revert to a more conservative lending policy—e.g., the FLBAs might limit the amount of money loaned on farmland to a lower percentage of the assessed value of the land, especially in high risk farming areas.

Farm lenders who served on credit task forces with the authors suggested that the present financial problems of farmers and the FCS are related to shortcomings in the financial management skills of farmers. This claim is supported by research results. To remedy this problem, additional efforts might be undertaken by the FCS, other private lenders, universities, and government agencies to improve the financial management skills of farmers, including their credit management abilities. In this regard, the Business Management in Agriculture project launched by the FCS and the Extension Service in 1986 to improve the profitability of U.S. agriculture by enhancing the financial management skills of farmers and agricultural lenders appears to be a promising step. If carried out effectively, such efforts could help farmers to make more effective use of credit and reduce the need for federal assistance to farm lenders in the future.

Possible Effects of a Farm Credit System - Farmers Home Administration Merger

What if the federal assistance and the changes described above fail to satisfactorily restore the financial health of the System? If such a situation materialized, one plan that might be considered is to merge all or part of the FCS with the Farmers Home Administration (FmHA). A unit consisting of many former FCS borrowers also could materialize without a FCS-FmHA merger if the most creditworthy FCS borrowers take their business to private lenders and the least creditworthy System borrowers find it necessary to borrow from an expanded FmHA. What have we learned from experiences with the FmHA that might be useful for structuring a combined organization or a larger FmHA which included large numbers of former FCS borrowers?

Origins of the Farmers Home Administration

While the FmHA is commonly known as the "lender of last resort" which provides farm mortgage and operating credit at below-market interest rates for farmers who cannot obtain credit from private lenders, the Agency (or its predecessors) has had far broader responsibilities from the early years of its existence. Beginning in 1937, a predecessor of the FmHA was given partial responsibility for making farm water system loans. Since then, the FmHA has had responsibility for making mortgage and operating loans to farmers and, in addition, has received responsibility for making rural housing loans, financing low rent apartment projects for senior citizens, water and waste disposal loan and grant programs, providing guarantees for rural business and industry loans, financing small-scale projects to produce synthetic fuels, making certain outdoor recreation loans, extending loans to limited resource farmers, making emergency disaster and economic emergency loans, and soil and water conservation loans to farmers.

<u>Experience With Farmers Home Administration Lending Programs</u>

The proliferation of FmHA programs has created problems for the Agency because the number of staff available to process and service loans has increased more slowly than loans outstanding. For example, the number of FmHA staff years for loan processing and servicing per million dollars of loans outstanding declined from 1.9 in 1963 to .18 in 1984 and the number of dollars for which one FmHA staff person is responsible increased from $.52 million in 1963 to $5.49 million in 1984. The impact of such changes on the Agency's ability to service its $26 billion in farm loans at a time when financially stressed FmHA farm borrowers required more than normal amounts of staff time is obvious.

As suggested by the statistics appearing in Table 4 (Tables at the end of chapter), the FmHA has experienced other problems. Recall that Harrington reported that about one-third of all U.S. farmers had debt-asset ratios exceeding 40 percent in 1985, which caused them to have serious or severe financial problems or to be technically insolvent. This contrasts sharply with the 83 percent of FmHA borrowers who had debt-asset ratios exceeding 40 percent in 1983-84 (Table 4).

The FmHA concedes that borrowers who are more than three years delinquent on loans are likely to default on the loans. As noted in Table 4, the dollar value of FmHA farm loans delinquent for more than three years totalled $5.5 billion in 1987. If one applies a 54 percent collateral recovery value suggested by Wilkinson for Federal Land Bank loans to the FmHA loans delinquent for more than three years, this produces estimated losses to the FmHA of $3.0 billion on these loans. In 1982-85, when the FmHA loans delinquent for more than three years as a percentage of the dollar amount of all delinquent loans increased sharply, the Farmers Home Administration allowed existing borrowers to obtain new financing without demonstrating an ability to repay previous, outstanding loans. This action may account for the sharp increase in the loan delinquency status during 1982-85. It is difficult to predict how much the losses might be on the remaining $20.5 billion in the FmHA's $26 billion farm loan portfolio. But, in view of the relatively high delinquency rate on the Agency's total farm loan portfolio, the losses could be large.

Among other things, the figures in Table 4 suggest

that the quality of the FmHA's farm loan portfolio has deteriorated in recent years. This deterioration doubtless reflects the worsening of the financial condition of the U.S. farm sector which occurred in the mid-1980s and administrative considerations which delayed foreclosures. The delays occurred for numerous reasons, including those associated with loan inquiries made by elected officials on behalf of constituents and the normal appeal process consisting of the informal meeting, the hearing, and the review used by borrowers when the FmHA proposes to take an adverse action such as foreclosure.

Legal challenges also have produced delays. For example, the Coleman v. Block class action suit filed on behalf of FmHA farm borrowers in 44 states in October 1983 produced a court order which specified that the Agency could not "terminate a farmer-borrower's living and operating allowance which was previously determined in the administration of any existing loan unless the borrower was given notice of the intended action, the right to contest the action and the right to apply for a deferral". This court order effectively produced a moratorium on FmHA loan foreclosures from February 1984 to February 1986.

Political actions have limited the discretion of the FmHA regarding the financial information that the agency could require borrowers to submit. In 1983, the Agency attempted to increase the documentation accompanying requests for FmHA loans by encouraging loan applicants to use certain Coordinated Financial Statements for Agriculture (balance sheets, income statements, and cash flow statements). Some FmHA borrowers objected to being required to use these financial statements, saying mainly that they were more complex than necessary. In response to requests by objectors, the Congress included a provision in the Food Security Act of 1985 specifying that the Coordinated Financial Statements would not be required by the FmHA.

Large prospective loan chargeoffs and the other problems experienced by the FmHA are perhaps predictable in times of farm financial stress, given the political pressures described above and the staff shortages faced by the Agency. Indeed, we might expect wholesale defaults when lenders and governments are not staunch about loan recovery and farm loans take on characteristics of income transfer programs. There is a substantial history of such results in developing countries. Thus, unless the merged organization is fundamentally different from the FmHA it is likely that a combined FCS-FmHA program would exhibit many of the problems that have affected the FmHA. We believe that the evidence raises questions about whether the FmHA would be an efficient substitute or support mechanism for the FCS if the 1987 federal assistance fails to work effectively.

A Modest Proposal

The FmHA might play a modestly larger role in buttressing the FCS in future years by providing more funds for loan guarantees. Expanded use might be made of plans to give System institutions guarantees not to exceed, say, 90 percent of any loss of principal and interest on a loan. The decision regarding whether to extend loans would be made mainly by the FCS. Increased use of loan guarantees would reduce the workload of FmHA personnel, remove the need for adverse actions by the Agency, and avoid some of the delays and administrative problems associated with such actions and give the FCS incentives to make sound loans since the guarantees would not cover all losses on a loan. This, of course, is a proposal which would work most effectively for moderately risky loans and does not address the problems of FCS borrowers who are presently near bankruptcy.

Summary and Implications

The FCS required federal help in 1985 and 1987 partly because the economic environment turned out to be harsher than expected. Although FCS officials might be criticized for taking an excessively sanguine view of economic growth prospects, the experience of the FCS during the mid-1980s mainly shows how difficult it is for firms to obtain accurate economic forecasts and the merits of strategies, mainly diversification of portfolio, which would enable the FCS to survive under a wider range of economic conditions. As a group, FCS borrowers were neither more nor less creditworthy than farmers who borrowed from commercial banks. Nonetheless, the financial problems of the FCS are more serious than those faced by the average commercial bank because of the heavier concentration of farm loans in the FCS portfolio and because a higher proportion of FCS loans have been made to large commercial farmers. The FCS also might be criticized for failure to accumulate adequate financial reserves during the early 1980s. The FLBAs charged relatively low interest rates during the early 1980s which helped the organization to gain market share but also produced problem loans and reserves which left the FCS inadequately prepared for the environment encountered in the mid-1980s.

The federal assistance for the FCS which was signed into law in December 1985 accomplished certain short-term objectives. In particular, the legislation helped to restore the confidence of financial markets in FCS securities and limit the number of creditworthy farm borrowers who withdrew their stock and took their business to other lenders during late 1985 and early 1986. The federal assistance for the FCS signed into law on January 6, 1988 provides financial assistance to the FCS, protects borrower stock and makes provisions for borrowers' rights, provides for System restructuring, establishes a Farm Credit Insurance Corporation and Fund, and creates a secondary market for agricultural real estate loans. However, restoration of the financial health of the FCS over the long-term may be difficult. The FCS may need to diversify its loan portfolio and help farmer-borrowers to manage credit more effectively if the financial health of the System is to show sustained improvement. A modest expansion of FmHA loan guarantees might be useful for buttressing the FCS in the future.

REFERENCES

1. David H. Harrington, "A Summary Report on the Financial Condition of Family-Size Commercial Farms," ERS-USDA, AB-492, March 1985.
2. Warren F. Lee, "The Farm Recession Continues: Some Prospects for Recovery," Socio-Economic Information, Department of Agricultural Economics and Rural Sociology, The Ohio State University, March 1, 1986.
3. Federal Farm Credit Banks Funding Corporation, "Farm Credit System Annual Information Statement-1987," New York, New York, March 11, 1988.
4. Farm Credit System, "Production Agriculture and Rural America in 1995," Project 1995, June 1984.
5. Charles Gratto, "The Economic Environment of the 1985 Farm Bill," Iowa State University's Ag. Policy Update, No. 13, September 13, 1985.
6. Emmanuel Melichar, "Farm Financial Experience and Agricultural Banking Experience," Statement presented to the Subcommittee on Economic Stabilization of the Committee on Banking, Finance, and Urban Affairs of the U.S. House of Representatives, Washington, D.C., October 23, 1985.

7. Emmanuel Melichar, "The Farm Credit Situation and the Status of Agricultural Banks," Paper presented at Twin Cities Agricultural Issues Round Table, St. Paul, Minnesota, April 24, 1986.
8. Capitol Publications, Blue Chip Economic Indicators, Vol. 9, No. 10, Arlington, Virginia, October 10, 1984.
9. Steven R. Guebert, ed., "Agricultural and Credit Outlook '86," Farm Credit Administration, McLean, Virginia, January 1986.
10. George D. Irwin, "The Farm Credit Crisis and Possible Solutions," Paper presented at the Louisiana Cotton Forum, Monroe, Louisiana, January 29, 1986.
11. James Johnson, S. Gabriel, and Kenneth Baum, "Aggregate Indicators of Financial Conditions in the Farm Sector: Some New Approaches and Insights," National Credit Conference Proceedings, NC-161, St. Louis, October 31-November 2, 1984.
12. U.S. Department of Agriculture, "Agricultural Finance Outlook and Situation Report," ERS AFO-26, Washington, D.C., March, 1986.
13. Lucy Huffman, Private communication regarding farm real estate debt market shares held by different lenders in 1980 and 1985, July 2, 1986.
14. Steven A. Loy, Letter describing the consolidation and mergers of the Production Credit Associations and Federal Land Bank Associations in the Fourth (Louisville, KY) Farm Credit District, January 17, 1986.
15. Steven R. Guebert, "Agricultural Situation Report," Farm Credit Administration, Washington, D.C., July 25, 1986.
16. Webster Communications Corporation, "The Agricultural Letter," Vol. 1, Letter 24, Washington, D.C., September 19, 1986.
17. Webster Communications Corporation, "The Agricultural Credit Letter," Vol. 1, Letter 19, Washington, D.C., July 4, 1986.
18. William J. Anderson, "The Farm Credit System, Analysis of Financial Condition," Statement before the Subcommittee on Conservation, Credit, and Rural Development of the Committee on Agriculture, U.S. House of Representatives, Washington, D.C., September 18, 1986.
19. Webster Communications Corporation, "The Agricultural Credit Letter," Vol. 2, Letter 4, Washington, D.C., November 21, 1986.

20. David A. Lins, "Financial Stress Among Farm Firms: Discussion," American Journal of Agricultural Economics, 67(5):1129-1130 (1985).
21. Gayle S. Willett, Private communication describing the Business Management in Agriculture project, October 27, 1986.
22. U.S. Department of Agriculture, "A Brief History of the Farmer's Home Administration," FmHA, Washington, D.C., February 1985.
23. Wilkinson, Donald E., Statement regarding the need for federal assistance for the Farm Credit System before the Subcommittee on Conservation and Credit of the House Committee on Agriculture, Washington, D.C., October 30, 1985.
24. Minnesota Legal Services Coalition, "Farmer's Guide to the Farmer's Home Administration," 3rd edition, January 1986.
25. Adams, Dale W., "The Conundrum of Successful Credit Projects in Floundering Rural Financial Markets," unpublished manuscript, April 1986.

Table 1. Forecasts Appearing in Project 1995 Study and Actual Values of Variables During 1973-81, 1982-84, 1985, 1986, and 1987.

Standard Forecasts in Project 1995 Study for 1985-1995	Actual Values for Variables During 1973-81	1982-84	1985	1986	1987
1. Generally steady, modest growth in real GNP, averaging 3.2%/year	2.5% /year	2.5% /year	3.0% /year	2.9% /year	2.9% /year
2. Unemployment drops to about 5% by 1995	6.7%	8.9%	7.2%	7.0%	6.2%
3. Prime interest rate drops to 8% or 9% if federal budgets are brought under control	10.7%	12.6%	9.9%	8.3%	8.2%
4. Domestic inflation rate averages 5% to 7%	9.1%	4.5%	3.6%	1.9%	3.7%
5. U.S. agricultural exports are expected to grow but not at the high rates of the 1970s	22.6% /year	-4.1% /year	-17.9% /year	-15.7% /year	6.1% /year
6. U.S. farm real estate debt rises from $121 billion in 1985 to $200 billion in 1995, average increase of 5.2%/year	13.1% /year	1.7% /year	-5.8% /year	-9.8% /year	-6.4 /year
7. U.S. farm nonreal estate debt rises from $106 billion in 1985 to $265 billion by 1995, average increase of 9.2%/year	14.0% /year	2.1% /year	-0.9% /year	-9.6% /year	-16.7% /year

Sources of actual values for GNP, unemployment, prime interest rate, and inflation: Council of Economic Advisers, Economic Report of the President, U.S. Government Printing Office: Washington, D.C. (1988) and Council of Economic Advisers, Economic Indicators, U.S. Government Printing Office: Washington, D.C., various issues (1986 and 1987).
Sources of actual values for U.S. agricultural exports: U.S. Department of Agriculture, Agricultural Statistics, U.S. Government Printing Office: Washington, D.C., various volumes (1975-1985) and U.S. Department of Agriculture, Outlook for U.S. Agricultural Exports, ERS-FAS, U.S. Department of Agriculture, December 2, 1986.
Sources of actual values for farm real estate and nonreal estate debt: Council of Economic Advisers, Economic Report of the President, U.S. Government Printing Office: Washington, D.C. (1988) and E. Melichar, "Farm Financial Experience and Agricultural Banking Experience" statement presented to the Subcommittee on Economic Stabilization of the Committee on Banking, Finance, and Urban Affairs of the U.S. House of Representatives, Washington, D.C., October 23, 1985 and "The Farm Credit Situation and the Status of Agricultural Banks," paper presented at Twin Cities Agricultural Issues Round Table, St. Paul, Minnesota, April 24, 1986.

Actual values for all variables are calendar year figures except those for agricultural exports which are fiscal year figures. The actual values for U.S. agricultural exports, farm real estate debt, and farm nonreal estate debt are year-over-year percentage changes.

Actual values for inflation reflected in the Consumer Price Index.

Based partly on preliminary debt figures for 1987.

Table 2. Chronology of Developments Preceding and Following Passage of Legislation Providing Federal Assistance for the Farm Credit System in December, 1985 and December 1987.

Date	Event
Feb. 1985	The Louisville FICB and FLB dismissed 85 of 380 employees to reduce costs.
May 1985	The interest rate spread between FCS six-month bonds and Treasury securities of the same maturity was about two basis points.
June 1985	FCS banks agreed to purchase $135 million of high risk or nonearning assets from the Spokane FICB.
July 1985	The worsening financial condition of U.S. farmers and implications of that development for the FCS appeared in a July 22, 1985 Wall Street Journal article.
Sept. 1985	FCS directors approved a $340 million assistance package to aid the Omaha FICB.
Sept. 1985	The Board of Directors of the 37 Farm Credit Banks agreed to form a special Legislative Committee to seek federal assistance to ensure the System's continued viability.
Oct. 1985	The interest rate spread between FCS six-month bonds and Treasury securities of comparable maturities rose to about 85 basis points.
Oct. 1985	D. Wilkinson, Governor of the FCA, sought federal assistance for the FCS, arguing that declining land values and other problems could put about $13 billion of FCS loans in nonaccrual status by the end of 1987. FCS officials made similar arguments in their requests for federal assistance.
Dec. 1985	FCS losses of $522 million were announced for the third quarter of 1985.
Dec. 23, 1985	Amendments to the Farm Credit Act of 1971 were signed into law to provide federal assistance for the FCS.
Feb. 1986	The FCS reported a $2.7 billion loss for 1985, a record loss for a U.S. financial institution which exceeded the previous record loss of $1.08 billion for the Continental Illinois Bank in 1984. Ten of the 12 Farm Credit Districts recorded losses (Springfield, Massachusetts and Baltimore were the two profitable districts) for 1985. The financial problems of the BCs were less severe than those of other components of the FCS.
Feb. 1986	H. Brent Beesley, Director of the Federal Savings and Loan Insurance Corporation during 1981-83, was named President and CEO of the Farm Credit Corporation of America which provides leadership and strategic planning for the FCS.
March 1986	Frank W. Naylor, U.S.D.A. Undersecretary, and Marvin Duncan, Senior Deputy Governor of the Farm Credit Administration, were selected for the positions of Chairman and member, respectively, of the new three-member FCA board. The U.S. Senate confirmed the appointments of Naylor and Duncan in May, 1986. J. Billington, a farmer and formerly a bank and PCA president, was confirmed by the U.S. Senate as the third member of the FCA board in October, 1986.

Table 2 (continued).

Date	Event
March 1986	Officials of the Amarillo, Texas PCA requested permission to withdraw from the FCS and convert the organization into a state-chartered mutual loan corporation partly to avoid sharing Association funds with financially weaker members of the FCS. FCA officials denied this request. Amarillo PCA officials challenged the FCA action in federal court, arguing that the action proposed by the PCA is not subject to FCA approval.
April 1986	Interest rate spread between a new $1.559 billion issue of FCS six-month bonds and Treasury issues of comparable maturities was 33 basis points.
May 1986	FCS losses for the first quarter of 1986 were reported as $206 million.
July 1986	FCA authorized a limited (0.5 percentage point) reduction in FCS interest rates to help stem losses of market share by the FCS.
July 1986	FCS losses of about $762 million were reported for the second quarter of 1986. The System's surplus declined from $3.4 billion at the end of 1985 to $2.4 billion on June 30, 1986.
August 1986	H.R. Macklin, former FLB, insurance, farm management and financial consulting firm official, was named President and CEO of the FCSCC.
July-Sept. 1986	A capital preservation agreement was activated to prevent capital stock impairments in the St. Paul, Omaha, Jackson, Louisville, St. Louis, and Wichita FLBs. Accrued contributions for the six FLBs totalled $415 million for July-September 1986. Assessments were initiated by the FCSCC which called upon stronger units of the FCS to invest $297 million in the FCSCC to help weaker System units. Legal challenges were filed against some of these assessments.
Oct. 1986	Legislation was signed into law which allowed FCS units to set interest rates without prior approval of the FCA and, with FCA approval, capitalize certain interest costs and a portion of their provisions for loan losses dating from July 1, 1986, and amortize such capitalized amounts for up to 20 years.
Nov. 1986	FCS losses of $560 million were reported for the third quarter of 1986. The System's surplus declined to $1.85 on September 30, 1986, down from $2.4 billion on June 30, 1986.
Feb. 1987	FCS losses of $1.9 billion were reported for 1986. The System's surplus declined to $1.45 billion on December 31, 1986. The financial problems of the BCs were again less severe than those of other components of the FCS.
March 1987	FCS asked Congress for $6 billion in direct assistance.
April 1987	FCS losses for the first quarter of 1987 were reported as $155 million.
July 1987	FCS losses for the second quarter of 1987 were reported as $46 million.
Nov. 1987	FCS profit for the third quarter of 1987 was reported as $4 million.
Jan. 6, 1988	The Agricultural Credit Act of 1987 was signed into law by the President of the United States.

Table 2 (continued).

Date	Event
Jan. 1988	The Farm Credit System Financial Assistance Board acquired the assets of and assumed the debts, obligations, contracts, and other liabilities of the Capital Corporation and the Capital Corporation was dissolved on January 21, 1988.
Feb. 1988	FCS reported net income of $180 million for the fourth quarter of 1987, resulting in a net loss for 1987 of $17 million.
Feb. 1988	On February 5, 1988, the Assistance Board returned to System institutions substantially all of the funds paid to the Capital Corporation under the assessment program. Some of the lawsuits against the Capital Corporation and the FCA by certain Banks and Associations were dismissed. For several others, joint plaintiff-defendant motions to dismiss are pending.
May 1988	The FCA appointed a receiver for the FLB of Jackson and its related FLBA to liquidate these institutions. The Assistance Board approved $5 million in assistance to permit the FLB of Jackson to retire eligible borrower stock at par value. In addition, six PCAs in liquidation in the Spokane and Omaha districts were granted $20.6 million to permit retirement of frozen borrower stock at par value.
May 1988	The Assistance board authorized aid to the Louisville FLB in the form of equity investments by the Financial Assistance Corporation totalling up to $90 million.
May 1988	FCS reported a net income of $165 million for the first quarter of 1988.
July 1988	Eight of the twelve Banks for Cooperatives voted to merge into a single national bank called the National Bank for Cooperatives. The four Banks for Cooperatives voting to remain independent are based in Springfield, Massachusetts; St. Paul, Minnesota; Jackson, Mississippi; and Spokane, Washington.

Source: Webster Communications Corporation, "Chronology of a Crisis: Agricultural Credit in 1985," McLean, Virginia, September 1985.
Source: Federal Farm Credit Banks Funding Corporation, Statistical Summary of Financings, 1985 and supplemental information on financings for 1986, New York.
Source: Farm Credit System, Reports to Investors, various reports, 1985, 1986 and 1987.
Source: Donald E. Wilkinson, Statement before the Subcommittee on Conservation and Credit of the House Committee on Agriculture, Washington, D.C., October 30, 1985.
Source: Webster Communications Corporation, "The Agricultural Credit Letter," Vol. 1, Letter 13, Washington, D.C., April 4, 1986.
Source: Webster Communications Corporation, "The Agricultural Credit Letter," Vol. 1, Letter 19, Washington, D.C., July 4, 1986.
Source: Farm Credit Banks of Louisville, "Venture," Vol. 3, No. 8, Louisville, KY, September 1986.
Source: Webster Communications Corporation, "The Agricultural Credit Letter," Vol. 2, Letter 4, Washington, D.C., November 21, 1986.

Table 2 (continued).

This legislation would allow the FCS to spread over 20 years the cost of refinancing $30.8 billion of high interest bonds and operating losses that exceeded 0.5 percent of outstanding loans.
Source: U.S.D.A., "Fundamental Changes Ahead for the Farm Credit System," Agricultural Outlook, December 1987.
Source: Farm Credit Banks Funding Corporation, Farm Credit System Annual Information Statement - 1987, New York, N.Y., March 11, 1988.
Source: Farm Credit Banks Funding Corporation, Farm Credit System Quarterly Information Statement - First Quarter 1988, New York, N.Y., May 10, 1988.
Source: Article in the Wall Street Journal, "Merger is Voted by 8 of the 12 Farm Banks," July 7, 1988.

Table 3. U.S. Farm Loans Provided by FCS and Other Lenders.

Debt Category and Lender	Amount of Loans Outstanding at Year End ($ billion)				Percent Change in Loans Outstanding
	1984	% of Total	1985	% of Total	from 1984 to 1985
Real Estate Debt	$111	99.9%	$107	100.1%	- 3.6%
Banks	10	9.0	11	10.3	+ 10.0
Federal Land Banks	49	44.1	45	42.1	- 8.2
Life Insurance Companies	12	10.8	12	11.2	0.0
Farmers Home Administration	10	9.0	11	10.3	+ 10.0
Individuals and Others	30	27.0	28	26.2	- 6.7
Nonreal Estate Debt	101	99.9	103	100.1	+ 2.0
Commodity Credit Corporation	9	8.9	18	17.5	+100.0
Banks	40	39.6	36	35.0	- 10.0
Production Credit Associations	18	17.8	14	13.6	- 22.2
Farmers Home Administration	16	15.8	18	17.5	+ 12.5
Individuals and Others	18	17.8	17	16.5	- 5.6
Total Farm Debt	$212		$210		- 0.9

Sources: E. Melichar, Statement on "Farm Financial Experience and Agricultural Banking Experience," presented to the Subcommittee on Economic Stabilization of the Committee on Banking, Finance, and Urban Affairs of the U.S. House of Representatives on October 23, 1985 and "The Farm Credit Situation and the Status of Agricultural Banks," paper presented at Twin Cities Agricultural Issues Round Table, St. Paul, Minnesota, April 24, 1986.
Totals do not sum to 100.0 because of rounding error.

Table 4. Selected Characteristics of Farmers Home Administration Farm Borrowers and the Farmers Home Administration Farm Loan Portfolio.

Characteristic	Findings
Debt-asset ratios and cash flow	The average debt-asset ratio for a sample of U.S. farmers who had FmHA loans in 1983-84 was 83 percent; about 83 percent of these FmHA borrowers had debt-asset ratios of 40 percent or more; and 85 percent of the borrowers in the sample had negative cash flows.
Loan delinquencies	The outstanding principal on delinquent FmHA farm loans (dollar amount) increased from 14 percent of total loan principal outstanding in 1976 to about 49 percent of loan principal outstanding in 1987. The amount of farm loans delinquent for more than three years expressed as a percentage of the total dollar amount of farm loan delinquencies increased from 21 percent in 1981 to 78 percent in 1987. The outstanding principal on farm loans delinquent for more than three years totalled $5 billion in 1987.

Source: U.S. General Accounting Office, Characteristics of Farmers Home Administration Borrowers, 1986.

Source: U.S. General Accounting Office, Farmers Home Administration: An Overview of Farmer Program Debt, Delinquencies and Loan Losses, June 30, 1987.

12

The Catholic Bishops' Pastoral Draft Statement on American Agriculture

Bishop Edward W. O'Rourke

On May 13, 1891, Pope Leo XIII published his famous encyclical "Rerum Novarum, the Condition of the Working Class." Pope Leo pointed out the several moral and value issues which are involved in economic policy, particularly as they relate to the working people in such economy. The American bishops are attempting to follow the precedent set by Pope Leo XIII, namely, to outline values and moral principles in order that the Catholic and other interested citizens of the United States might participate in our economy and in our political system with such values and moral principles clearly in mind. My task in this paper is to analyze the chief features of one part of the American bishops' pastoral statement, namely, "Food and Agriculture."[1] First, I would like to briefly summarize a few thoughts found in the second chapter of the Pastoral as a background to the observations with regard to food and agriculture. The second chapter of this Pastoral is entitled "The Christian Vision of Economic Life." Reference is made to the fact that each human being is a part of God's creation and made to His image and likeness. As a consequence, "As such every human being possesses an inalienable dignity which stamps human existence prior to any division into races or nations and prior to human labor and human achievement (Gn. 4-11)."[2] After reviewing biblical references to wealth and riches, we conclude: "Such perspectives provide a basis for what today is called the 'preferential option for the poor.'"[3] With regard to love and solidarity we read: "The commandments to love God with all one's heart and to love one's

The author thanks the United States Catholic Conference for permission to use excerpts from the Third Draft--Economic Justic for All: Catholic Social Teaching and the U.S. Economy, Copyright © 1986, United States Catholic Conference, Washington, D.C. A final version of the Pastoral Statement was approved by the bishops in November 1986.

neighbor as oneself are the heart and soul of Christian morality. These commands point out the path toward true human fulfillment and happiness. They are not arbitrary restrictions on human freedom."[4] In addition to justice, as usually conceived by our contemporaries, we point out the implications of related types of justice, namely, social and distributive justice with the following observations.

> Social justice implies that persons have an obligation to be active and productive participants in the life of society and that society has a duty to enable them to participate in this way.[5] Distributive justice requires that the allocation of income, wealth, and power in society be evaluated in light of its effects on persons whose basic material needs are unmet.[6]

Among the chief duties of a government is to promote the common good. We find the following observations about that subject:

> The common good demands justice for all, the protection of the human rights of all.[7] The obligation to provide justice for all means that the poor have the single most urgent claim on the conscience of the nation.[8] Increasing active participation in economic life by those who are presently excluded or vulnerable is a high social priority.[9]

We bishops come out squarely in defense of the right of workers to organize unions:

> The Church fully supports the right of workers to form unions or other associations to secure their rights to fair wages and working conditions.[10]

Also pertinent to our discussion is Chapter IV of the pastoral entitled "A New American Experiment: Partnership for the Public Good."

> We believe that completing the unfinished business of the American experiment will call for imaginative new forms of cooperation and partnership among those whose daily work is the source of the prosperity and justice of the nation. The United States prides

> itself on both its competitive sense of initiative and its spirit of teamwork. Today a greater spirit of partnership and teamwork is needed; competition alone will not do the job.[11]

Throughout the rest of that chapter various suggestions are made about self-help in general, the forming of cooperatives, and profit-sharing, ownership and other forms of cooperation within particular industries. Historically, agriculture has been one of those industries in which cooperatives have abounded. We will return to this theme as we now give some detail with regard to the proposal made by the bishops with regard to food and agriculture.

The following notations from this section on food and agriculture summarize the main themes which we bishops are proposing.

> The fundamental test of an economy is its ability to meet the essential human needs of society in an equitable fashion. Food, water, and energy are essential to life; and their abundance in the United States has tended to make us complacent about them. But these goods--the foundation of God's gift of life--are too crucial to be taken for granted. Our Christian faith calls us to look constantly at God's creative and sustaining action and to evaluate our own cooperation with the Creator in using the resources of the earth to meet human needs.[12]

> No aspect of this concern is more pressing than the nation's food system. Just as food is a unique, life-sustaining commodity, farming is a special vocation. As pastors we feel the distress of many farm people who are threatened with bankruptcy or are facing foreclosure.[13]

Noting that the Pre-emption Acts of the early 19th century and the Homestead Act of 1862 brought about a wide distribution of ownership of farm land, recent trends have caused a great concentration of such ownership in the hands of fewer and fewer people. Noting also the U.S. government's policy to keep food costs low, we read,

> While low food prices benefit consumers, who are left with additional income to spend on other goods, such pricing policies put pressure on farmers to increase

> output and hold down costs. This has lead them to replace human labor with cheaper energy, expand farm size to employ new technologies favoring larger scale operations, neglect soil and water conservation, underpay farmworkers, and oppose farmworker unionization.[14]
>
> The situation of racial minorities in the U.S. food system is a matter of special pastoral concern. They are largely excluded from significant participation in the farm economy. Despite the agrarian heritage of so many Hispanics, for example, they operate only a minute fraction of America's farms. Black-owned farms, at one time a significant resource for black participation in the economy, have been disappearing at a dramatic rate in recent years, a trend that the U.S. Commission on Civil Rights has warned 'can only serve to further diminish the stake of blacks in the social order and reinforce their skepticism regarding the concept of equality under the law.'[15]

Among the "guidelines for action," the following might be noted.

> First, moderate-sized farms operated by families on a full-time basis should be preserved and their economic viability protected. Similarly, small farms and part-time farming, particularly in areas close to cities, should be encouraged[16]....
>
> Second, the opportunity to engage in farming should be protected as a valuable form of work. At a time when unemployment in the country is already too high, any unnecessary increase in unemployed people, however small, should be avoided. Farm unemployment leads to further rural unemployment as rural businesses lose their customers and close. The loss of people from the land also entails the loss of expertise in farm and land management and creates a need for retraining and relocating another group of displaced workers[17]....
>
> Third, effective stewardship of our natural resources should be a central consideration in any measures regarding U.S. agriculture. Such stewardship is a contribution to the common good that is difficult to assess in purely economic terms, because it involves

the care of resources entrusted to us by our creator for the benefit of all.[18]

Also we find the following suggested "Policies and Actions."

A prime consideration in all agriculture trade and food assistance policies should be the contribution our nation can make to global food security. This means continuing and increasing food aid without depressing Third World markets or using food as a weapon in international politics[19] .... The current crisis calls for special measures to assist otherwise viable family farms that are threatened with bankruptcy or foreclosing. Operators of such farms should have access to emergency credit and programs of debt restructuring[20] .... Established federal farm programs, whose benefits now go disproportionately to the largest farmers, should be reassessed for their long-term effects on the structure of agriculture. Income-support programs that help farmers according to the amount of food they produce or the number of acres they farm should be subject to limits that ensure a fair income to farm families and restrict participation to producers who genuinely need such income assistance. There should also be a strict ceiling on price-support payments which assist farmers in times of falling prices, so that benefits go to farmers of moderate or small size[21] .... Although it is often assumed that farms must grow in size in order to make the most efficient and productive use of sophisticated and costly technologies, numerous studies have shown that medium-sized commercial farms achieve most of the technical cost efficiencies available in agriculture today. We therefore recommend that the research and extension resources of the federal government and the nation's land-grant colleges and universities be redirected toward improving the productivity of small and medium-sized farms[22] .... Since soil and water conservation, like other efforts to protect the environment, are contributions to the good of the whole society, it is appropriate for the public to bear a share of the cost of these practices and to set standards for environmental protection. Government should therefore encourage farmers to adopt more conserving practices and distribute the

> costs of this conservation more broadly[23].... The ever present temptation to individualism and greed must be countered by a determined movement toward solidarity in the farm community. It is possible to approach farming in a cooperative way, working with other farmers in the purchase of supplies and equipment and in the marketing of produce. It is not necessary for every farmer to be in competition against every other farmer.... Farmers also must end their opposition to farmworker unionization efforts. Farmworkers have a legitimate right to belong to unions of their choice and to bargain collectively for just wages and working conditions. In pursuing that right they are protecting the value of labor in agriculture, a protection that also applies to farmers who devote their own labor to their farm operations.[24]

This section on food and agriculture is concluded with the following observations.

> The U.S. food system is an integral part of the larger economy of the nation and the world. The very nature of agricultural enterprise and the family farm traditions of this country have kept it a highly competitive sector with a widely dispersed ownership of the most fundamental input to production, the land. That competitive, diverse structure, proven to be a dependable source of nutritious and affordable food for this country and millions of people in other parts of the world, is now threatened. The food necessary for life, the land and water resources needed to produce that food, and the way of life of the people who make the land productive are at risk. Catholic social and ethical traditions attribute moral significance to each of these. Our response to the present situation should reflect a sensitivity to that moral significance, a determination that the United States will play its appropriate role in meeting global food needs, and a commitment to bequeath to future generations an enhanced natural environment and the same ready access to the necessities of life that most of us enjoy today.[25]

These are a few insights into the bishops' proposed pastoral on Catholic social teaching and U.S. economy with particular emphasis on the section on food and agriculture. We hope to further refine this statement in a meeting in June of this year at Collegeville, Minnesota, and hopefully be prepared to adopt a Pastoral on this subject in our General Meeting in Washington, D.C., during November of this year.

Endnotes

1. Pastoral Letter on Catholic Social Teaching and the U.S. Economy, Second Draft, Washington, D.C., National Conference of Catholic Bishops, October 7, 1985, chap. IIIC.
2. Pastoral Letter, para. 38, p. 11.
3. Ibid., para. 59, p. 18.
4. Ibid., para. 69, p. 21.
5. Ibid., para. 75, p. 23.
6. Ibid., para. 76, p. 23.
7. Ibid., para. 87, p. 26.
8. Ibid., para. 88, p. 26.
9. Ibid., para. 93, p. 28.
10. Ibid., para. 103, p. 30.
11. Ibid., para. 284, p. 77.
12. Ibid., para. 213, p. 58.
13. Ibid., para. 214, p. 58.
14. Ibid., para. 217, p. 58.
15. Ibid., para. 223, p. 60.
16. Ibid., para. 229, p. 61.
17. Ibid., para. 232, p. 62.
18. Ibid., para. 234, pp. 62-63.
19. Ibid., para. 236, p. 63.
20. Ibid., para. 238, p. 63.
21. Ibid., para. 239, pp. 63-64.
22. Ibid., para. 241, p. 64.
23. Ibid., para. 242, p. 64.
24. Ibid., para. 244, p. 64-65.
25. Ibid., para. 246, p. 65.

13

Response to Bishop O'Rourke

George Horwich

The Catholic bishops' blueprint for American agriculture, as summarized by Bishop O'Rourke, embodies a pastoral vision of farm life containing many appealing elements--at least to nonfarmers. It does not, however, have much to do with economic reality; nor is it likely to be achieved without being mandated by governmental authorities. Accordingly, an economist who usually favors free-market outcomes is unlikely to find much in the blueprint that he can support.

What is missing in the bishops' pastoral letter is an understanding of general economic principles and of how economic forces have shaped the agricultural sector both recently and over the long sweep of American history. Against this backdrop of economic development and the incomparably successful record of American food production and farm income, I, for one, would hesitate to tamper with the sector, even in time of crisis, or to alter its institutions in any major way.

But the bishops are less restrained. They want to reduce reliance on competitive forces, preserve moderate-sized family farms and encourage small ones, slow the transfer of farm land to nonfarm uses, increase farmworker unionization, expand farm ownership by racial minorities, and end U.S. policies which they believe are generally designed to keep food costs low but simultaneously inhibit progress toward these goals.

I will turn to some of the specifics of farm experience in a moment. Let me say first that I found Bishop O'Rourke's opening reference to the "preferential option for the poor," a central theme of the pastoral letter, somewhat misplaced in a discussion of American agriculture. The bishops want to promote the welfare of the poor not only in the narrow sense of subsidizing their incomes, but in the broader sense of elevating their social status and influence. I know of no social policies capable of accomplishing this larger goal, but even if they existed, their relevance to American farmers in current financial difficulty is practically nil. As I interpret the letter, the unmistakable object of the bishops' preferential option are the inner city underclass and the unschooled, unskilled rural poor--"those marginalized...persons who have 'no voice and no choice.'" This does not sound much like the owners of U.S. farm land, including those whose speculative investments of the last decade have turned sour with the decline of farmland prices and the rise of overseas agricultural competition. These owners, many of them in dire straits, are far from voiceless or without choice. Although low interest government loans fueled the land boom, it is hard to argue that these farmers, solidly middle class and articulate, have a preferential social claim that any unsuccessful investor in any sector of the economy does not also have.

To take a longer view, the history of this republic has been one of a continuing exodus from the land. In 1776, at its founding, 90 or 95 percent of the population were engaged in farming. Today the number is less than 3 percent and continues to fall. Although government intervened at many points in this process and at times influenced its pace, government is not responsible for it. The underlying cause is the growth of agricultural productivity, reflecting increased mechanization, improved fertilizers, innovations in crop and animal breeding, and advances in general farming techniques. All of these, in a competitive free market setting, enabled a drastically shrinking fraction of the population to feed an ever increasing number of people. In light of these forces, the movement off the land was and is unavoidable and necessary if consumers are to enjoy the benefits of low cost and plentiful food supply.

One cannot conclude, therefore, that any current abandonment of the land or transfer to other uses is unreflective of the long-term economic trend and would not have occurred sooner or later even in the absence of any particular perturbation of agricultural or real estate values. Given the simultaneous growth of technology and the size of the market (U.S. real per capita income has risen at least 12 or 13-fold since 1776), it would also be surprising if the average size of farm, as dictated by economic factors, did not increase from time to time. I challenge Bishop O'Rourke's statement that most of the technical cost efficiencies in agriculture are achieved by medium-sized commercial farms. The word medium is, of course, relative, and whatever it may be today, it will be different tomorrow.

Clearly the bishops are observing a dynamic process through static eyes. They miss the fact that if farmers fail to adopt more efficient techniques (such as increasing the average size of farm when warranted) or if resources (land and labor) fail to leave agriculture when new technology reduces their required input, society experiences no net economic gain.

The bishops' suggestion, moreover, that government policies are providing cheap food at the cost of the farmers' wellbeing ignores the specific form of these policies. Direct government subsidies are indeed substantial, accounting today for 40 or 50 percent of total farm income. This in itself has the effect of keeping more farmers, including a limited number of small ones, in farming than would the free market. But the subsidies are primarily price supports that raise the cost of food to all consumers.

The bishops rightly celebrate the values and the virtue of the family farm and its role in our economic development. As long as the small or moderate-sized family farm plays an economic role, it can indeed serve as a model of independent, industrious, family oriented activity. But to try to freeze the farm of given dimensions in opposition to economic forces will fail not only on the economic front, but in the values realm as well. When the farm of given size is the product of economically based initiative and hard work, that activity and its underlying behavioral norms are likely be highly valued in this society. But when that same farm is propped up and supported essentially by governmental subsidies and decrees that run counter to the forces of technology and consumer choice, the behavioral pattern associated with it is likely to be enervating and incompatible with widely shared American values.

I would also like to dissent from the unspoken implication that the values of suburbia, to which most surplus farm resources gravitate, are in general of lower quality or status than those of the family farm. With all their problems, suburban, and even much of urban, America still provide a salutary lifestyle for the masses. The decline of the frontier has led to a transfer of the American self-help ethic first to an urban, and then to a suburban setting. We should not ignore this development or apologize for it.

An increase in the number of black, Hispanic, and other minority landowners, strongly urged by the bishops, is, like the small family farm, something that natural market forces have not provided. The determinants of ethnic occupational choice and location are, to say the least, complex. In this century the primary movement of blacks has been away from the rural south, where blacks typically held poverty-level sharecropper status in agriculture. This migration is credited, in a recent study by the Rand Corporation, with a major role in raising the national ratio of black to white male wages from 43 to 73 percent between 1940 and 1980.[1]

To suggest that blacks might simultaneously or shortly thereafter leave their sharecropper status for managerial and ownership status on the land ignores not only the shrinking importance of land, but some crucial links in the typical course of human economic progress, as well. The bishops assume implicitly that black (and Hispanic) ownership can be facilitated by public policy measures. But the Rand study found that schooling was the only policy variable influencing the black/white wage ratio. Preferential programs, such as affirmative action, had no effect on the relative wage level, though they did appear to open up a number of occupations to black males. I am not sure, however, that the long-term impact of such programs is truly beneficial. There is much evidence that they are not.[2]

The case for unionizing farmworkers, also encouraged by the bishops, likewise fades under empirical scrutiny. Unions in American history have doubtless played a role in developing worker morale and group identification. As an economic force, however, they typically raise wages of workers who are organized 5 to 15 percent while reducing the wages of those who are not organized.[3] There is no evidence that unions raise the general level of wages in an industry except by reducing the size of the industry. Unionism, as an economic force, is not something that American agriculture particularly needs at this time.

FOOTNOTES

1 James P. Smith and Finis R. Welch, Closing the Gap: Forty Years of Economic Progress for Blacks (Santa Monica: Rand, 1986), R-3330-DOL.

2 See Thomas Sowell, The Economics and Politics of Race: An International Prospective (New York: Quill, 1983), pp. 131-32, 200-201, 253-54.

3 H. Gregg Lewis, Union Relative Wage Effects: A Survey (Chicago: University of Chicago Press, 1986).

PART THREE

Synthesis of Papers

14

The Family Farm: Shall We Freeze It in Place or Free It to Adjust?

Earl L. Butz

Much discussion of the U.S. farm crisis today focuses on the family farm, its role in the historical development of America, its contribution to our moral and social values, and its status in contemporary rural America.

Some view with alarm the economic and technological forces that threaten the very existence of the family farm as they knew it in their youth. They see only negatives for America as change engulfs rural life. Others see a new and different rural America emerging, with a new and economically stronger family farm dominating the agricultural community. Which pattern is best depends, I presume, upon the social and economic prejudices of the analyst.

I grew up on a family farm in northern Indiana. It was a little larger than most in that community. It consisted of 160 acres, 120 of which were tillable. (The typical farm was a family unit of 80 or 120 acres.) We had 25 beef cattle, 8 sows, and 100 hens. Motive power was provided by 5 horses (two teams with an extra horse when a 3-horse hitch was needed). This was a two-man farm, with extra help for haying and harvest. Dad and the live-in hired man constituted the labor force (mother tended the chickens) until I was 14 years old, when I replaced the hired man. When I left to attend Purdue University at the age of 19, my next younger brother replaced me in the farm labor force.

Four years later when I had graduated from Purdue, my younger brother left the farm to enter the university. I returned to the farm as the farm operator; Dad was by now manager of the local Farm Bureau Cooperative. It was still a family farm, pretty much unchanged from three decades earlier when my grandfather first acquired it. The farm family furnished the capital (including debt-free land), the management, and the labor. We produced and processed most of our own food. Barnyard manure made expenditure for fertilizer unnecessary. Clover in the rotation provided the nitrogen for next year's corn crop. Home-grown hay and oats for the horses held fuel purchases to a minimum. The farm didn't generate a big cash flow, but we didn't need much. We were pretty self sufficient. So were all the neighboring farmers. We all lived on and operated family farms. None of us had much extra money. We all envied those life's amenities that our city cousins could afford and enjoy.

After one year as operator, I left the family farm to pursue graduate study at Purdue, never to return to production agriculture. My two younger brothers and my youngest sister and her husband likewise used college education as an escape route. One sister and her husband lived on the farm and operated it for some years. However, they took off-farm employment and rented that 160 acres to a neighbor family farmer, who was enlarging his operation as a part owner, part renter. The neighbor now lives on a 120 acre family farm which he owns, and rents some half-dozen other family farms in the neighborhood which are owned by one-time farmers. The new family farmer in the community now operates some 800 acres. But he is still a family farmer.

I am glad I left production agriculture and did not try to continue on that family farm of 160 acres. I am glad I can afford two automobiles and an air conditioned home and that we could afford to educate our two sons and take some nice family vacation trips as they were growing up.

What pleases me most is that the neighbor who remained on the land and now operates a family farm of 800 acres can afford the same things I can. If I, or my brothers, had remained on our 160 acre family farm, and if our five neighbors had continued to operate their 80 to 120 acre farms, none of us could have afforded any of the amenities we now enjoy unless we had a substantial off-farm income.

This personal story is far from unique; it is a small part of that great farm exodus so many of us lament today. Yet, we didn't think it wrong when we left production agriculture. No one forced us to remain on the family farm where we grew up nor did anyone force us to leave. In most cases, there was no federal program to "keep us down on the farm." We were mobile. We recognized that agriculture was in the midst of a technological and scientific revolution that was changing forever the face of the family farm. The farm was getting larger. Capital was displacing labor as the primary production input. Adequate cash flow was needed to meet the increasing cost of off-farm produced inputs. As labor productivity increased, added volume was needed to utilize the available labor supply (the operator and/or family labor). This requirement for more volume meant more acres or more intensive enterprises on given acres, or in many cases, off-farm employment.

This shift has been going on for decades. As we look back, we think the process, on the whole, was good. As we look ahead, we are apprehensive. Why should we fear a continuation of what we have regarded as good? A decade hence we'll look back on the decade of the 80's and pronounce it good, as well.

A dozen years ago I sat in Brezhnev's office in Moscow discussing his agriculture and food situation. With remarkable frankness, Brezhnev said, "I have 40 percent of my people on the land. I can't begin to produce consumer goods in abundance for my people until we learn to feed them more efficiently."

I sat there thinking: "Brezhnev, you have put your finger precisely on the Achilles Heel of the Russian economy. Forty percent on the land! Your country right now is where my country was when I was born in 1909."

In my lifetime, the United States has moved from 40 percent of our people on the land to 2.2 percent today. Our total population has grown from 90 million to 240 million, an increase of 2 1/2 times. The number of people fed per worker on U.S. farms has grown from about 5 to well over 75—over 100 per farm worker on our full-time commercial farms.

The relative cost of eating in this country has dropped markedly. When I was born, it took two-fifths to one-half of our income to feed ourselves. Now we spend just a mite over 15 percent of disposable income on food, and that includes all the built-in maid service we purchase simultaneously and the nearly one-third of our total meals taken outside the home. That leaves some 85 percent of our take-home pay for all those goods and services that make possible the highest and most widely distributed level of affluence among all the nations of the earth.

As I heard Brezhnev complain that 40 percent of his people were on the land, I was thankful that at my birth, when the United States likewise had 40 percent on the land, some politician, some rural sociologist, or some rural preacher didn't succeed in freezing the pattern and stopping off-farm migration.

By the same token, we should oppose current efforts to halt off-farm migration. The adjustment is not complete. Obviously, at some point we reach an irreducible minimum number of workers in production agriculture. But we are not there yet. There are more efficiencies to be attained and more unit-cost reductions to be sought.

It is true that U.S. agriculture currently has plenty of individual distress cases. They dominate the rural news. The media emphasizes failure, foreclosure, off-farm migration. The public impression is that agriculture, as an industry, is about to go down the tube, that the industry is sick and that everyone in it is sick. Nothing is further from the truth. There are success stories among today's farmers--they just don't make the news. Nearly half of our farmers have no debt. Net cash farm income in 1986, including government payments to farmers, will be near a record.

But there are failures on the farm front. Agriculture, however, is not unique in that respect. As I drive down the street, I see some closed filling stations, I see some closed supermarkets, I see some afternoon newspapers that didn't make it, I see many small business ventures that couldn't cut the ice.

This is the American system. It's a system of risk and reward, of success and failure. In our free enterprise system you can't have the successes without the failures.

Some day I am going to develop Butz's law of economics: "In any industry, if government sets out to guarantee everyone against failure, then it also removes the possibility of success beyond mediocrity."

Governments never level up; they only level down. Big is bad, small is good. Governments level toward smallness. They move toward minimum entitlements for the small producer and maximum allotments for the large producer. They level toward inefficiency in the industry. They level toward whatever it takes to keep the least efficient in business. To do that they have to put a cap on the most efficient. That's the only way government knows how to ration opportunity.

The economic definition and the political concept of a family farm are two different animals. The economic definition is a farm large enough to provide a decent living for the farm family. The political concept is a farm small enough so that you starve slowly.

Somebody has defined capitalism as the unequal distribution of plenty, socialism as the equal distribution of scarcity. When government gets involved, the equal distribution of scarcity is the route most commonly traveled. And when we do that in any industry, we become high-cost producers. In time, we become low-income producers.

We need a farm program that provides flexibility, that recognizes differences in ability, and acknowledges that perhaps we should not all succeed in the business.

How do we address the problem? How do we again unleash the tremendous productive capacity of our nation's bread basket? How do we redirect world grain production to areas of greatest efficiency and lowest unit cost, both at home and abroad? How do we re-establish our competitive edge in the world's export markets? How do we lift the onus of production controls from our farmers? How do we restore the hope of expanding markets and rising income?

It is increasingly obvious that the answer does not lie in the course we have been following. That path leads inevitably to expanded competition abroad, to loss of foreign markets, to curtailed production at home, to the proliferation of governmental controls, and to reduced income.

What we have been doing obviously doesn't work. In the long run, farm income can never be enhanced and rural welfare can never be achieved through a program of restriction, rising unit costs, unrealistic pricing, market withdrawal, expanded governmental controls, or growing dependence on income transfer payments from the U.S. treasury.

The answer must be found in other directions. Key words will be: efficiency, lower unit costs, competitive pricing in export markets, favorable macro policies in the nonagricultural sectors of government, loan rates at market clearing levels, a safety net under agriculture at a level that quits pretending that a government bin is a market, a signal to our competition abroad that "the honeymoon is over."

There is a discipline in the market place that works. Let's give it a try.

15

Subsidies and Support

Gerald J. Lynch

There is an overwhelming presence of government intervention in the agricultural sector of the economy, both domestically and internationally. The first question that comes to my mind is, why is there such a large role for government in the agricultural sector of the economy? Many of the papers addressed that question.

Terry Roe suggests that government policies toward agriculture often have a goal not connected to agriculture, but use agriculture as a means of attaining that goal. For example, one measure that indicates the lack of development in a country is the percentage of the labor forced involved in agriculture. Seeing that the developed countries have experienced migration off the farms and into the manufacturing sector, developing countries attempt to emulate that migration at a time when the economy is not ready for it. The result is often that agriculture is burdened with resource-distorting taxes to subsidize an infant manufacturing industry that is far from ready to be competitive on a worldwide basis. Roe points out that not all of the tax on agriculture is siphoned off into another sector. Much of the revenue goes into public investments that benefit agriculture. However, even with agriculture reaping some gains from this public sector investment, the form of intervention is such that improper signals are sent regarding the allocation of resources.

One of the ways that developing governments attempt to increase manufacturing in their countries is, in effect, to raise the real wage of manufacturing workers through lower food prices. Roe notes that in many developing countries, food expenditures comprise up to 60 percent of the total budget. If food prices are suddenly lowered, the real purchasing power of workers rises and labor resources will flood into the manufacturing sector. It seems an ingenious plan but the implementation refuses to recognize the existence of an upward sloping supply curve.

Here is an area where, as an economist not familiar with the agricultural economic world, I learned an astonishing fact. In developing countries, it is common practice to hold down the price of agricultural products so that the real wages of urban manufacturing workers can rise. A disastrous result of this policy is that agricultural production is held down because there is a reduced return to farming. The food shortages in so many developing countries are artificial constructs of some government-operated food-marketing board. I am reminded of a similar discovery I recently made about why Washington's troops had such a hard winter at Valley Forge. In an effort to hold down the prices of basic supplies for the Army, the Continental Congress put price controls on basic food commodities and firewood. The result was a shortage of both foodstuffs and firewood throughout the colonies that winter, not just in Valley Forge. Whenever shortages of this type result, people spend an inordinate amount of time and effort on exchange instead of production. The end result might well be lower production in both agriculture and manufacturing.

I am a great believer that a market system will send out proper allocative signals in most instances. If a country is not currently an exporter of manufactured products, that is because someone can produce those products more cheaply abroad and, even with transportation costs, sell them more cheaply in the home country. While a manufacturing sector seems to be a good source of wealth, forcing a country into manufacturing may reduce overall wealth.

I was struck by a similar thought as I read Alain de Janvry's paper on land reform. There are two arguments as to why land reform will benefit the developing countries and, in particular, the rural poor. One is that larger farms are, overall, less productive and total wealth will increase if those farms are broken up into smaller farms. The other argument is that the ownership of small farms is a means of escape from poverty for many of the rural poor. The second argument is one that I have little quarrel with. De Janvry presents statistics to support the idea that land reform is a form of income redistribution. If a country wants to engage in that form of income redistribution, then so be it.

I have a more difficult time with the idea that land reform will increase overall factor productivity because smaller farms will be more productive than larger ones. If that were the case, smaller farms would have emerged over time in a profit maximizing setting rather than large farms. De Janvry notes that many of the conditions that weaken the overall increase in factor productivity associated with land reform are present and that land reform may well not increase productivity. Once again, I would trust what has emerged naturally. Larger farms must be more productive because that is what has prevailed.

There are, of course, situations, such as those involving externalities and public goods, where government intervention is generally accepted. These situations, however, are not the type discussed above with regard to agriculture in developing countries and they are not the arguments for agricultural assistance from the government in the United States.

The next question that comes to my mind then is, absent from the arguments made above for government intervention into the agricultural sector in the developing countries, what role should government play? Edward Schuh's paper that a proper macroeconomic setting is a sine qua non, not only for agriculture, but for all business. Unfortunately, that setting has not been provided for the American farmer recently. Schuh and Kelley White both noted that the fluctuating value of the dollar has created a great deal of uncertainty for farmers selling their goods abroad. Over about a 36 month period from late 1981 through early 1985, the value of the dollar in international markets rose by some 70 percent against an index of all other currencies. That means that a dollar cost foreign countries about 70 percent more and thus, American made goods, agriculture included, cost about 70 percent more. No merchant would raise his prices by that much that fast if he had the power to control the process, but firms selling abroad do not have that power. It is important to note, however, that the farmer is not the only person hurt by this type of exchange instability.

That brings up a third, final, and perhaps most emotionally charged question. Should the farmer have special protection from the government? I want to address this question primarily from a domestic point of view.

There is no question that the farmer does receive

special protection. We think of protection for the farmer coming primarily in the form of price supports, but they are not the only form of government assistance to agriculture. There is also a large amount of support that goes to farmers through the Farmers Home Administration (FmHA). Let's look at the size of each of these just through the previous (1981-86) farm bill. I want to look at the previous farm bill because, as you will see, the estimates of what a bill will cost and the actual costs often diverge by huge amounts. Looking at the projected costs on the 1986 farm bill may give us little clue of the actual cost.

When the farm bill was passed in 1981 it was assumed that the cost of agricultural subsidies would total $11 billion. This assumption was based on continuing inflation that would keep the price of farm products high. In one of the ironies of government intervention, the grain support subsidy was a program that got more expensive as the rate of inflation fell. When farm prices did not increase as rapidly as had been predicted, the difference between the market price and the guaranteed price grew, as did the size of the subsidy. Thus, a program that was supposed to cost $11 billion over five years, wound up costing $53 billion, with over $14 billion spent in 1985 alone.

The fact that farm prices are supported at all shows a different attitude towards farmers than towards other segments of the economy. The idea of the subsidy is that the real price of farm products should not fall. Yet, as the real price of energy fell from the post World War II ear through the early 1970s we did not see a rush of aid to the oil companies. As the real price of electronic goods, e.g., home computers, falls, we do not see legislation to prop up the price. In this regard, farmers are simply treated differently.

The second largest subsidy to farmers are the loans made by the FmHA. FmHA assistance is available to farmers who have been turned down by commercial banks or production-credit associations. In other words, farmers who were deemed not creditworthy by the market received loans at below-market interest rates that amounted to a subsidy of about $3 billion dollars. There is no question that farmers receive special treatment. That treatment cost a total of $21 billion dollars in 1985. The interesting question is why are farmers singled out?

I don't really have the answer and I can't claim that singling the farmer out is necessarily wrong, but I perceive that most of the special treatment is aimed at maintaining the family farm. Even though there are often claims that the farm sector is shrinking in terms of the number of people employed, that may simply be a result of not realistically evaluating what constitutes farm work. Admittedly, the number of people breaking the soil has diminished. Yet agricultural industries, from farming to the retailing of farm products, still constitutes the largest proprietary sector of the American economy, accounting for 20 percent of GNP and employing more people than the steel and automobile industries combined. The son of the production farmer may now be an agricultural engineer designing computerized combines, but he is still in agriculture. That is why I think what bothers people most is the passing of the family farm. If we were to find out that the large agribusiness farms were going under and that the family farm was flourishing, I doubt that we would see the flood of sentiment that exists over the passing of the family farm. Few economic endeavors evoke the lyrical passages that farming does.

We did not lament the passing of the Mom and Pop grocery store--at least, we did not lament it enough to subsidize it so that it could compete with supermarkets. Sometimes when I drive to Indianapolis I take the old highway 52 instead of going by the Interstate. As I drive along, I am always struck by the sight of the roadside motels often constructed of cinder block with faded neon signs that always advertise vacancies if they are open at all, motels that look like Norman Bates is the proprietor. Those motels were "done in" by the building of the Interstate. The traffic no longer passed their door and the weary travelers no longer stopped. Yet, when that happened there were no reams of paper filled with elegiac prose about their passing. There were no bills put before Congress that would force travelers to stay there. The motel owners, the Mom and Pop grocery store owners, and a host of other small businesses that fell by the wayside as the structure of the economy changed did not have a Willy Nelson to sing a lament for their passing.

I don't know why that is, but when I hear discussions about fairness to the farmer, I think of the fairness of the existence of farm subsidies and the absence of motel subsidies. Bishop O'Rourke noted that many farmers were surveyed by researchers for the Catholic Bishops before the pastoral letter was issued. They found that family farmers love what they are doing and want to maintain their lifestyle. Thus, the Bishops conclude, it is our duty to maintain their dignity. As I read the Bishop's paper, I thought of the work of Studs Terkel. Studs interviews everybody and does it about as well as anybody on this planet. A few years ago he wrote a book called Working that was based on these interviews. He found out that a large number of people working in factory jobs absolutely hate their jobs and wish there was some way out of them. And what I wonder is this: If the Bishops think we should subsidize the farmer to keep him in his job because he loves it, should we also do something to help those factory workers because they hate their jobs?

I do not claim to have the prescience to know why we support farmers and not others, but I do believe we do it out of emotionalism and not because it is economically expedient.

About the Contributors

Freddie L. Barnard is assistant professor of agricultural economics at Purdue University.

Earl L. Butz is dean emeritus of agriculture at Purdue University and a former United States Secretary of Agriculture.

Alain de Janvry is professor and chairman of the department of agricultural and resource economics at the University of California, Berkeley.

William D. Dobson is professor of agricultural economics at Purdue University.

William E. Foster is assistant professor of economics at North Carolina State University.

George Horwich is professor of economics and Burton D. Morgan Professor for the Study of Private Enterprise at Purdue University and a collaborating scientist of the Oak Ridge National Laboratory.

D. Gale Johnson is the Eliakim Hastings Moore Distinguished Service Professor of Economics at the University of Chicago.

Melvyn Krauss is professor of economics at New York University and a senior fellow of the Hoover Institution at Stanford University.

Gerald J. Lynch is associate professor of economics and associate director of the Center for Economic Education at Purdue University.

John W. Mellor is director of the International Food Policy Research Institute in Washington, D.C.

Edward W. O'Rourke is the Catholic Bishop of Peoria, Ill.

Don Paarlberg is professor emeritus of agricultural economics at Purdue University.

Gordon C. Rausser is the Robert Gordon Sproul and Class of 1934 Chair Professor of Agricultural and Resource Economics at the University of California, Berkeley.

Terry L. Roe is professor of agricultural and applied economics at the University of Minnesota.

G. Edward Schuh is dean of the Hubert H. Humphrey Institute of Public Affairs, professor of public affairs, and professor of agricultural and applied economics at the University of Minnesota.

Mathew Shane is deputy director of the Agricultural and Trade Analysis Division of the Economic Research Service of the U.S. Department of Agriculture.

T. Kelley White is director of the Agricultural and Trade Analysis Division of the Economic Research Service of the U.S. Department of Agriculture.

For Product Safety Concerns and Information please contact our EU representative GPSR@taylorandfrancis.com
Taylor & Francis Verlag GmbH, Kaufingerstraße 24, 80331 München, Germany

www.ingramcontent.com/pod-product-compliance
Ingram Content Group UK Ltd.
Pitfield, Milton Keynes, MK11 3LW, UK
UKHW022205190926
475702UK00003B/212

* 9 7 8 0 3 6 7 0 1 3 0 0 4 *